The social impact of oil

Robert Moore
Professor of Sociology, Aberdeen University

The social impact of oil

The case of Peterhead

Routledge & Kegan Paul
London, Boston, Melbourne and Henley

First published in 1982
by Routledge & Kegan Paul Ltd
39 Store Street, London WC1E 7DD,
9 Park Street, Boston, Mass. 02108, USA,
296 Beaconsfield Parade, Middle Park,
Melbourne, 3206, Australia, and
Broadway House, Newtown Road,
Henley-on-Thames, Oxon RG9 1EN

Set in Press Roman by
Hope Services, Abingdon, Oxon
and printed in Great Britain by
Billing & Sons Ltd,
Guildford, London, Oxford and Worcester

Library of Congress Cataloging in Publication Data

Moore, Robert (Robert E.)
The social impact of oil.
Includes index.
1. Petroleum industry and trade – Scotland – Peterhead
(Grampian)   2. Petroleum industry and trade – Environmental
aspects – Scotland – Peterhead (Grampian)   3. Fish trade –
Scotland – Peterhead (Grampian)   4. Peterhead (Grampian) –
Economic conditions.   5. Peterhead (Grampian) – Social
conditions.   I. Title.
HD9571.7.S36M66   306'.3   81-19984

ISBN 0-7100-0903-8   AACR2

For Lindy, David and Heloïse

Contents

Acknowledgments

The research upon which this book is based was funded by the Social Science Research Council and was intended to be part of a much larger programme of research.* My special thanks are due to Professor Fred Martin, the Chairman of the SSRC's North Sea Oil Panel, whose practical help and moral support enabled us to sustain our efforts in studying the social impact of North Sea oil from 1976 to 1978. Whilst I concentrated most of my efforts in Peterhead, Dan Shapiro was studying migrant labour further afield, but his work analysing labour records in Peterhead with the help of Barry Deas and Gordon Philip was invaluable. Maggie Wilson helped me in the library and in Peterhead and produced the first complete party-political analysis of local government in the north of Scotland.

Colleagues in the University of Aberdeen were especially helpful; John Sewel was an indispensable source of local knowledge and he and Frank Bealey freely gave me access to the typescript of their Peterhead book,† for which I am very grateful. Members of the Department of Sociology tolerated my absences and preoccupation whilst Jean Smith kept the day-to-day administration of a large department running smoothly when I was away. Linda McHardy and Ruby Heath undertook the typing and retyping of my text. I thank them for working wonders.

Much of my work was concerned with planning, land use and public enquiries. With this and much else I received a great deal of help from

---

*See R. Moore, 'Sociologists Not At Work' in G. Littlejohn et al. (eds) *Power and the State*, Croom Helm, 1978, pp. 267–302

†Frank Bealey and John Sewel, *The Politics of Independence*: *a Study of a Scottish Town*, Aberdeen University Press, 1981.

Maurice O'Carroll of Grampian Region Planning Department and Gordon
Mann, Director of Planning for Banff and Buchan District, and their
staff. I am extremely grateful to them. The staff of Companies House
and the Register of Sasines in Edinburgh were very kind and went out
of their way to be helpful. I would also like to thank Mr Cameron of the
ESA for his assistance in Peterhead; Mr Ayton, Director, and Mr Fyvie
of the Housing Department, the director of the social work division and
many other officers of local authority and state agencies. The Grampian
Project of the Scottish Council of Social Services was promoting community participation in Peterhead before I arrived and they gave me
information and help with this research.

The people of Peterhead had been the objects of many studies from
foreign television crews to English college students. They had been
written about, broadcast, filmed, photographed and yet still they found
time to talk to me and show me many kindnesses. It would be unreal-
istic for me to thank them all by name but I must mention collectively,
the ex-burgh councillors and officers; the Harbour Trustees; local trades
unions and the officers of the Professional and Businessmen's Association.

Many individual colleagues acted as sounding-boards for ideas and I
thank them. But research is only possible when it seems worth while,
and this depends in large part upon the intellectual climate in which one
works. The Department of Sociology in this university is a very good
place to be, not least because it provides the necessary intellectual stim-
ulus and support in a friendly atmosphere, so I am deeply indebted to
all my colleagues here. My family not only had to put up with my absences in Peterhead but with being deprived of the car thereby — I
would like to thank them for being so tolerant.

What I have to say in this book will not please everyone, but those
who are not pleased should not blame those whom I have thanked. If
they are in some cases the same persons I make no apologies, and my
thanks to them are none the less.

Robert Moore
Aberdeen, August 1978

Introduction

The thirty miles of countryside between Aberdeen and Peterhead could easily win a prize for the dullest scenery in Britain, save in the autumn. Then the rolling land is an amazing patchwork of reds, gold and green under a brilliant blue sky. The scenery is beautiful to the point of being a hazard to road traffic. The immediate approach to Peterhead is by a coast road past one of the bleakest prisons in Britain. Granite and barbed wire stand on a bluff facing into the merciless northeasterly winds. But having rounded the prison, one sees Peterhead across the flat expanse of the Harbour of Refuge. Old Peterhead is built of pink granite and on a clear day in evening or winter sun the town looks like a Canaletto painting. Once in the town one discovers the fishing harbours where in 1971 it was possible to watch a seal playing in the water amongst the fishing boats at Sabbath's rest.

In 1978, before the prison one passes the chimney of Boddam power station which can be seen from Aberdeen when the top is not obscured by cloud. Once past the prison the scene is transformed. Ahead and to the left a seemingly vast housing development extends three-quarters of a mile southwestward from the old town. The Harbour of Refuge (renamed the Bay Harbour) may be dominated by a giant oil-rig, the top of its derrick may be seen from anywhere in the town. In October, for example, a rig was careened alongside the breakwater for repairs and its massive weight and height tilted drunkenly and dramatically. At each side of the harbour are jetties filled with red, orange, yellow and brown supply vessels and backed by warehouses and hoppers containing drilling 'mud' and oil-well cement. At night rigs, jetties and boats are bright under floodlights. Something has happened to Peterhead, and that something is oil.

Map 1   Location of Peterhead

Peterhead is geographically remote and receives no casual visitors. In a BBC news bulletin Fraserburgh (near to Peterhead) was shown on a map as near Edinburgh, the equivalent of locating Leicester in Southsea. The town was probably only known to seafarers and to connoisseurs of Protestant sectarianism. Seen from the North Sea oil and gas fields, however, Peterhead is the nearest point of land. Given its proximity to the fields, it was inevitable that Peterhead and its harbour would be used and that oil and gas would make landfall nearby. Terminals, gas separation plants, transport and material storage all use space and land was going to be used upon which to build such facilities. Furthermore if any 'downstream' activity based upon the processing of oil and gas products was to take place then even more land would be needed. Not surprisingly, perhaps, land speculation is the subject of the first chapter of this book.

Oil actually came then in a variety of ways; the servicing of offshore activity from the harbours, the construction of landfall installations for gas and oil to the north and south of the town, and the promise of petro-chemical industries which would have transformed the whole economic base of the town. All these activities have taken place in about five years, from 1973 to 1978. The question has often been asked, 'What is happening to Peterhead?' but it is a question that is asked from a number of quite different viewpoints.

For the Peterheadian it is a question about the quality of his or her life. Quite dramatic and visible changes that they did not ask for have been thrust upon them. The streets seemed to fill with Italians and Spaniards, and then empty again. The prices of houses seemed to rise suddenly. There were more jobs and rising wages. The streets became littered, the pavements broken and heavy traffic made it dangerous to cross the road. There were underlying anxieties; marriages seemed to be breaking up, there were more serious crimes before the courts, the schools were overcrowded. Were the good times to be brief and Peter-head soon cast back into economic decline and unemployment? These anxieties were backed by a feeling that the local people could do very little about the events which were willed by powerful economic and political forces far beyond their control, and then even their burgh council was abolished in 1975 and the nominal seat of local political power removed from the town. Of the changes that came about, how many were attributable to fishing, to oil or to other causes and would they be Good or Bad for Peterhead?

Secondly, the planners, social workers and development agencies wanted to know what was happening to the local labour market; were locals being taken from their present jobs by incomers, were they de-skilled or given further training, were their wages higher or lower than before? If workers were coming in would they settle with their families or move on, and in either case what housing demand would they create? Would Peterhead's population grow temporarily or permanently and what services would be needed to cope with the changed population and what provision would the caring agencies have to make to meet the special needs of the times?

These are two quite legitimate and valid sets of questions, the first being mainly qualitative and concerned with moral issues and questions of morale. The second set of questions is more technical and demands quantifiable answers. But there is also a problem of perspective. The north and northeast of Scotland contains 14 per cent of the population of Scotland and 1.4 per cent of the UK population. In the small towns heavy unemployment is measured in hundreds. This has to be contrasted with the chronic unemployment of many thousands and the industrial dereliction of the populous Midlands. Is it unduly idealistic to look only at the effects upon a small population whilst ignoring the benefits to millions? One altruistic Peterheadian suggested that in 'the national interest' even a 'St Kilda solution' (evacuation of the whole population) was acceptable if it saved the British economy. Drastic as this may seem and given that Peterheadians are neither a rare nor a migratory species, such a policy could be seen as acceptable for the greater good in a programme either of accelerated capital accumulation or of a socialist reconstruction. None the less not only might the people of the northeast wish to debate the issue, but we might also ask what kind of *national* interest can make the interests of 14,500 people disposable. It is suggested later that it may be misleading to think in terms of an homogeneous 'national' interest. But crucially the fact remains that what happens in the northeast is relatively unimportant to most Scots or to the UK population at large — and this is part of the problem we have to address.

The sociologist is also concerned with these questions and finds himself at times drawn into the orbit of both groups of questioners and pressed to provide simple answers — after all, he is trained to answer the questions and has time to do research upon them, doesn't he? At the same time the sociologist asks himself questions that arise out of his own theoretical orientations, that is to say out of his understanding of how institutions, social relations, ideas and beliefs are connected (although not always obviously to the layman). So in observing Peterhead he is not looking at a town that is as unique as it is to a Peterheadian nor at a location which simply has problems of planning and policy, as an administrator might see it. Theoretical issues are raised by events in Peterhead and it is the questions raised by these to which the sociologist primarily applies himself, believing that whilst he may not immediately satisfy either the Peterheadian or the planner he will be able to offer them a wider and more general perspective upon the events that interest them and help them to a fuller understanding thereby. This, at least, is the ideal. How far this research falls short of the ideal is for the reader to judge.

A brief general discussion of theoretical problems is reserved for the Conclusions but I hope the general reader (if there is such a reader) will not skip the discussion because not only does theory enable us firstly to understand and secondly transcend our immediate experience but it actually becomes part of the understanding of those who control our present and future. In other words, politicians and administrators have theories and we need both to understand and to be able to criticise these theories in order to understand and criticise policy. Sociological theory may be abstract but its consequences are not.

Without pre-empting the concluding discussion we can say that explanations of the social change, and their consequences, similar to those taking place in Peterhead have centred around ideas like industrialisation and development. But the idea of industrialisation does not seem appropriate to areas that are already part of an industrial society or actually in industrial decline. Furthermore, the idea itself is often quite a muddled one based upon the history of nineteenth-century Manchester and Adam Smith's political economy. The condition of areas on the geographical margins of industrial society has to be explained in part by the relations between the periphery and the industrial 'centre'. This has led us directly to development theory (or theories) because explanations of development or underdevelopment in third-world countries have needed to refer to the relations between these 'peripheral' countries and the industrialised nations of the 'centre'. Failure to develop, in popular discussion often equated with failure to industrialise, is not the result of a natural process but the outcome of the political and economic subordination of the third-world country to the industrialised nations. Underdevelopment is not a state but a process. Given the internally uneven development of industrialised nations, it seemed reasonable to try out similar theories at home. This has not been altogether successful because the analogy between nation-states and regions of a nation does not stand up to the test of use. More importantly, this kind of analysis emphasises geographical or spatial inequalities which may be more accurately understood as social inequalities — but this really is to anticipate the Conclusions.

Suffice it to say that I have used ideas of development and underdevelopment as a way into understanding Peterhead. I am sure my professional colleagues will be touched by my naïveté when I translate this theoretical stance into an innocent question and some simple preliminary answers. The question is: how will we recognise development or underdevelopment in Peterhead, by what features of the social

changes will the sociologist be able to identify the symptoms of one or
the other? We will look at four aspects of life in Peterhead: economic
changes, the behaviour of the local elite, the mobilisation of labour and
the availability and distribution of social resources. If Peterhead is
undergoing development we would expect to find the following: existing businesses expanding either directly into oil-related activity or into
new markets created by improved wage levels. New or expanded manufacturing or service industries will find a potential for self-sustained
growth independent of the oil 'boom'; such enterprises might be in
boatbuilding and repair, offshore maintenance or the retail trade.

The traditional elite will be displaced by a new business elite and a
new middle class of scientists, technicians and administrators concerned
with research, development and administration in the expanding local
economy.

Labour we would expect to find mobilised to campaign for more
development and higher wages. Local and incoming workers would
transcend their differences of interest to act as a class against employers
locally and nationally. The Labour party would gain members and activists and thereby improve its electoral chances. It should be noted in
this context that development might give rise to new or increased conflicts in the community.

Social resources would expand to meet demand. For example housing
both publicly and privately built would be available to meet population
increase and homelessness and to improve housing standards. Schools
would be built and enlarged and manned with qualified teachers.

Underdevelopment would be characterised by the weakening or
elimination of existing businesses which would be forced to reduce their
activity, close down or move away because of the loss of labour or increased wages. The local elite or a new elite would monopolise economic
opportunities by acting on behalf of incoming companies. Their position
would be reinforced and the dependence of others upon them increased
because of their powers of patronage derived from the oil industry.
Incoming and local labour might be insulated from one another, possibly
by the use of camps. Incoming companies (possibly backed by the state
in collaboration with national union leadership) would use well-tried
techniques to reduce the relative power and autonomy of those workers
who are unionised locally.

On the question of social resources we might find either of two rather
different situations; (1) overstretched resources leading to long waiting
lists for housing, price escalation and emigration, overcrowded schools

using temporary accommodation with high teacher turnover induced by poor working conditions and falling pass rates in public examinations or (2) large-scale public investment unused, houses standing empty, expanded schools with declining rolls, resources tied up in buildings and overstaffing, redundancies amongst teachers.

In the chapters that follow we will describe events and conditions in Peterhead in order to see which set of circumstances has more nearly come about and therefore whether Peterhead is developing or underdeveloping as a result of oil.

Underpinning nearly all these considerations are the political questions of whether anything could be different and the extent to which the people of Peterhead are able to control the outcome of events. Peterhead is a very small speck on the map and it has a tiny population. Its interests feature not at all in the policies of the major corporations who pursue profit world-wide. The interests of Peterhead are also very slight in the consideration of the state which sees itself as pursuing national regeneration through oil exploitation. Conflicts of interest over resources and their use take place outside Peterhead, what happens in Peterhead is the result of bargaining, negotiation, power struggles between companies and between capital and the state. The most powerful interests operating in Peterhead can be controlled at the level of the state and the state in turn adopts policies which have a profound effect upon Peterhead. We will constantly have to take this into account when looking at wage levels, taxation, the uses of harbours and land in and around Peterhead and ask if Peterhead's ability to control its own destiny has been enhanced or reduced by recent developments. To bring this point home and to see how Peterhead may only be one aspect of a company's strategy — Peterhead as *part* of a plan, nor the whole of it — we look in Chapter 1 at land speculation. We take one company which operated throughout the north of Scotland with Peterhead as only one of its undertakings. Then when we consider the company in Peterhead we see how its operations relate to similar companies and other interests. Having then seen Peterhead as only one part of a national company's range of operations and policy options we thereafter confine our discussion to the town itself and see how the wider world of politics and economics impinges upon the 'Bluetoon'.

The second chapter looks at the way in which some of this land was used when gas from the North Sea made its first impact upon the town. Especially important in this chapter is the question of how planners could plan in a state of uncertainty and technical ignorance. Chapter 3 examines the way in which the people of Peterhead themselves sought to exercise some sort of control over the events that were happening to them, with particular reference to two important planning enquiries. In Chapter 4 there is a brief analysis of the use and management of the Peterhead harbours, the local resource over which there was considerable conflict. Chapter 5 turns to certains aspects of the structure of the Peterhead community, especially its recent political history and the interests that were served by the (now abolished) burgh council. This chapter ends with a discussion of the ways in which the policies of the state impinge upon Peterhead, both to its detriment and advantage, calling into question the whole notion of the town as an autonomous 'community' which might have controlled the impact of North Sea developments. The sixth chapter attempts, at some length, to answer the question 'Well, what actually happened in Peterhead – what was the effect of oil?' The answer is framed mainly in terms of effects on employment and the local labour market; housing and social resources; 'social problems'; entrepreneurial activities and the organisation of trade unions. The final chapter returns to more general questions of the processes under way in Peterhead and how we understand them.

1 Speculation

Peterhead has two assets which made it attractive to the oil industries; its location and its harbour. The northeasterly corner of Scotland offered the best mainland landfall for the Frigg and Brent gas fields 380 and 520 kilometres to the northeast and the closest landfall for the Forties oil field, 185 kilometres to the east. The gas fields are closer to the Shetlands and to Norway, but a Shetland landfall would have been most unprofitable as the gas would have had to be liquefied and shipped out again. Norway was thought to be impossible because of the deep ocean trench off the Norwegian coast — across which not even Norway's share of Frigg could be carried. The coast north of Peterhead offered an ideal landfall (gently sloping sand) and possible sites for preparing the gas for immediate supply to the national gas grid. The Forties oil was brought directly to Cruden Bay (13 km south of Peterhead) and this in 1976 became the first British offshore oil field to come on stream.

Peterhead's proximity to oil and gas fields did not make its harbour an obvious choice given the space and services available in Aberdeen 42 kilometres to the south. But Aberdeen harbours soon became crowded and Peterhead was chosen both by private entrepreneurs and the state as a suitable base for servicing offshore vessels and installations in the development stage of the offshore industry. We will look in detail at developments in the harbours below (Chapter 4) but it is useful at this stage to set Peterhead in a slightly wider context and to look at the local activities of land speculators to see what effect they had on subsequent policy choices made in the Peterhead area.

The development of offshore oil has an onshore effect in raising the price of land because it is self-evidently the case that facilities will be needed for the building, servicing and processing of oil and gas. The rise in land prices is a source of profit which may accrue to a householder who sells part of his garden to a house-builder or to a firm which buys large tracts of land, or takes options upon them in anticipation of later sales. 'Pure' speculation is a wholly non-productive and parasitic operation which entails buying cheaply and selling dear and thus, in effect, depriving the original owner of profit. The profit is based on the accident or good fortune of prices rising as a consequence of decisions made by others. Private *development* is not so wholly parasitic, because it involves at least a minimum input of physical services and expert knowledge, plus the ability to bring together the services with wholly or partially serviced land and the businesses which need these facilities, in order to build industrial plant or housing.

The distinction between pure speculation and development is one recognised by speculators and developers. The latter are careful to avoid being thought the former. There were successful developers in the north and northeast during the period 1970-77, but other alleged developers appear to have been speculators either by intention or through bad judgment. Some, presumably by bad judgment, simply made losses.

Land speculation was first reported in Peterhead in 1720 when the burgh was purchased by the York Building Company.[1] The burgh had originally been developed by the Earls Marischal from a grant of James VI in 1587. But in 1715 the Tenth Earl Marischal sided with the Old Pretender and his lands were forfeit. In the interregnum, between 1715 and 1720, the feuars (tenants holding a *feu* from a *feu superior*) encroached upon the land around them. This encroachment gave rise to disputes that were not finally to be settled until 1879. In 1728 the Merchant Maiden Hospital of Edinburgh purchased the burgh and improvements to the harbour were set in train which brought prosperity to the town.

In '45 Peterhead was for the Young Pretender and a chance to get rid of the 'robber merchants from Edinburgh'.[2] This phrase was to be echoed in Parliament over 229 years later when the 'Edinburgh mafia' were accused of speculation in the region.

Two features of the land transactions and development proposals in the northeast in the 1970s are notable: firstly, land passed out of local ownership, if it had not done so previously. Ownership might pass to locally registered companies, like Nordport of Lerwick in Shetland, for example, but invariably they are the subsidiaries of companies based outside the region (Onshore Investments Ltd) or outside Scotland altogether (Brandts Second Nominees Ltd). The land was usually

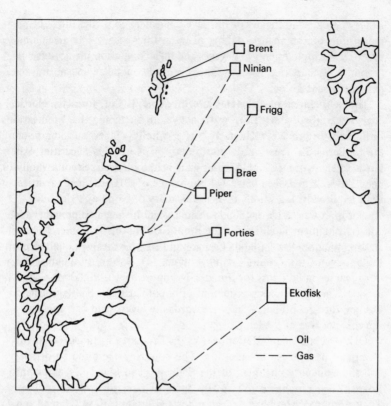

Map 2   Peterhead and the North Sea, 1977

purchased with a standard security from a national bank or finance company. None the less these subsidiary companies seem to have built a 'local dimension' into their operations and by so doing they have been able to capitalise upon expert local knowledge. The usual way of achieving this has been to involve local law firms in their activities and in some cases forming joint enterprises with local investors and property companies through the agency of the law firms.

Secondly, developers presented plans made on a scale which might seem more appropriate to central or regional government (for example, Cromarty Firth Development Company's plans for 'Nairn to Dornoch', *The Scotsman*, 7 March 1973). It is possible that these plans made the running for the region's planners and we will suggest below that this may have been the case in Peterhead. Perhaps more importantly the grand —

if not grandiose — schemes of the developers help construct the public's perception of the changes taking place in their locality. In areas long beset by unemployment, low wages and high emigration the promise of 'boom' and economic regeneration is one with popular appeal but, to date, little substance.

The subsidiaries of OIL (Onshore Investments Ltd, formerly North Sea Assets (Scotland) Ltd) were active in Shetland, the Cromarty Firth and Peterhead. OIL itself was a wholly owned subsidiary of the St Bernard's Trusts Ltd, a subsidiary of North Sea Securities. All but a £1 share of OIL's £1M shares were held by Brandts Second Nominees Ltd and in December 1974 the whole of OIL was made over to Brandts. Brandts is a wholly owned subsidiary of Grindley's Holdings.

Nordport was OIL's Shetland subsidiary and it sought to provide base and transhipment facilities for the Brent field. In this and other ways it clashed with Shetland Islands Council and its activities were halted by the passage of the Zetland County Council Act. But political defeat did not mean economic loss for Brandts. Nordport paid £290,000 for 1400 acres of land. 925 acres was compulsorily purchased by Shetland Island Council for £2.2 million — in other words they were paid over ten times the cost per acre of the land.

OIL's second manifestation was as the Cromarty Firth Development Company based near Invergordon. The company recruited an official and the ex-deputy chairman of the Highlands and Islands Development Board. They purchased over 3,500 acres of land for over £3 million. They proposed a terminal and an industrial estate (MIDAS), an air terminal and a motel plus housing developments which brought them into conflict with the local authority. By August 1973 the company was bankrupt.

In the Buchan area OIL was represented by Peterhead and Fraserburgh Estates (PFE). According to Rosie[3] a local lawyer, John (Jock) Smith, had been involved in setting up the company. This would be consistent with Smith's other activities in the property field. PFE had a share capital of £10,000 and its last published accounts were qualified by the accountants.

On 1 November 1974, Mr Barclay concluded some transactions that may not have altogether pleased him. On 30 September he had sold Wellington Place Farm to PFE, but on 1 November he disponed* his property to Peterhead and Fraserburgh Estates Ltd *and* to Scanitro.

*The Scottish legal term for conveyancing land.

Scanitro was the Scandinavian consortium who intended to build an ammonia plant in Peterhead. Between contracting the sale and signing the disposition PFE had sold 102.84 acres to Scanitro. Barclay received £180,744 for 223.3 acres and Peterhead and Fraserburgh Estates £400,000 for the 102 acres in the northwest of the farm. Barclay also sold Upperton and Wellbank Farms, an acreage of 197.7 for £269,580 to Peterhead and Fraserburgh Estates. He had tried to buy up land in the Peterhead area with a view to profitable sales. PFE had none the less plainly deprived him of some of his potential profit.

In a booklet 'Why Peterhead?', produced for an exhibition in 1973, the Company said the following about itself:

> Peterhead and Fraserburgh Estates Limited has been incorporated to help meet the need for back-up land behind the massive developments now underway by Aberdeen Service Company and Arunta. The company has acquired control of land situated in the immediate vicinity of the town and is developing substantial areas of land close to the harbour as industrial and residential estates to support the two oil bases. Advance warehouses, factories, and offices will be built so that incoming industries' total requirements can be met 'off-the-shelf'. Special areas of land are being set aside for pipe storage, craneage, and individual developments.
>
> A joint company has been formed with Arunta to develop the areas beyond the immediate vicinity of Keith Inch. Talks are also being held with other development companies in the area with a view to formulating a comprehensive development programme.

The company seemed to believe that formulating 'a comprehensive development programme' was a function of private enterprise. To assist in its planning Peterhead and Fraserburgh Estates had a programmer who had previously been Aberdeen County Planning Officer before leaving to join a partnership who were planning consultants to North Sea Assets.[4]

The planning was not public: 'Onshore's land holdings in Peterhead are being developed (sic) in conjunction with other land holdings by an associated company on behalf of Peterhead and Fraserburgh Estates. *Behind-the-scenes* work is actively under way and it is hoped detailed plans will be announced by the beginning of the new year' (my emphasis).

The 'associated company' working behind-the-scenes was Arunta

Properties (Scotland) Ltd, a joint undertaking with Arunta International, which revealed its plans in March 1974. The scheme was described in a publication, *Peterhead North Sea Terminus*, and entailed an ambitious development to the southwest of the town. The company held the freehold of 402 acres (Wellback, Upperton and Wellington Place — Barclay having at this time agreed to sell to Brandts who were acting for Arunta Properties (Scotland) Ltd) and about 110 acres of this were zoned for industrial development. The company, however, expressed a 'justifiable hope' that 'further industrial development could reasonably be expected to take place' on parts of the remainder (§ 3.3). The project, which appeared to have been proposed on the basis of careful market research and land surveys, included the provision of: pipe storage, industrial warehousing, lorry park, car showrooms and commercial vehicle service area, office block, flats, hotel, recreational developments, a heliport and landscaping (§ 4.4). The second phase of the plan also included an ammonia plant.

The financing of the project was calculated over 11½ years and based on maximum loan finance of £3.1 million to be repaid at about 16 per cent per annum. Total interest payments were reckoned to be £2.45 million, but in the period it was estimated that the company would have financed an investment of about £4.2 million from the profits after taxation, which would yield 25 per cent pre-tax. There is a slightly ambiguous note to the finances (§ 4.5) suggesting that some operators might prefer to do their own construction and that a direct sale of 10 acres might cover this and, moreover, that an early sale would help finance the company's investments. Elsewhere these sales are described as 'presently proposed for liquidity reasons'. Plainly the *sale* of land at a profit was an integral part of the development plans of Arunta Properties (Scotland) Ltd and the failure to sell land may, in part, account for the apparent inactivity of the company in Peterhead. The Arunta interest became dormant and Barclay finally disposed of the land to PFE, as we have seen, and they concluded the Scanitro deal.

One of OIL's backers, North Sea Securities, had a 50 per cent share in the Buchan Development Company and had made it a £40,000 loan. This is the last of the OIL ventures studied in Peterhead and it is an interesting one. According to the *Scotsman* of 20 September 1974 the Buchan Development Company was a joint venture between PFE and Site Preparations and according to the records in Companies House the collaboration was short-lived.

Who were Site Preparations Ltd? On 12 February 1973 this company purchased parts of Invernettie Lodge and Damhead Farm for £150,000 (110 acres) and part of Newmills of Sandford (35 acres) for £30,000. On 22 March they purchased the 71 acres of Whitehill Farm for £106,285. In September, November and December they made further purchases: the Meethill part of Invernettie (104 acres) for £138,750; Denend Farm (26 acres) and Farm of Buckie (98 acres) for £186,656 each and Middle Grange (13.9 acres) for £41,769. Thus in one year Site Preparations acquired 458 acres of land for £654,000. About 250 acres of the 458 were zoned land. Site Preparations thus held about half of all the land zoned for industrial development. Site Preparations in fact held options on much of this land in 1972 when the Secretary of State agreed to the re-zoning of land around Peterhead for industrial and commercial use.

It is too simple to see this only as a case of a firm acquiring zoned land. The activity and interest of Site Preparations was a material factor in the production of the Peterhead Town Plan. This is not to say that the plan was entirely led by the developers; the land was suitable for development, especially if development was intended close to the Harbour of Refuge. The extent and boundaries of the zoning would, however, be influenced by the presence of a willing developer.

Soon after the land purchases boards appeared at the roadsides; on the A952 by the new ring road, Property and Estates Management (Edinburgh) announced a site for a hotel and conference centre; opposite the prison, land for luxury housing; by Dales Cottages they advertised a site for an abattoir and meat factory and below Grange Farm, on the west road, a location for executive housing. Property and Estate Management (Edinburgh) was a subsidiary of Site Preparations, who held 85 per cent of its shares. In its most recent accounts the turnover was shown to be £20,619 of which £20,600 was fees for work done for the parent company. In 1977 the viability of this company depended on attempts to liquidate Site Preparations.

Site Preparations Ltd was founded in 1967 with a capital of £100 and it is one of the 21 companies run by J. M. and I. M. Dickson who occupied a small office in Barrhead (Glasgow).* The money for their

*According to *The Scotsman*, 30 September 1974, 13 of these firms have never traded.

purchases was raised from Old Broad Street Securities and, for Middle Grange, Crédit Lyonnais.

It was, presumably, the function of Property and Estate Management to sell the idea of building abattoirs and executive housing to developers who would then have the land and the expertise of the Dicksons at their disposal. It was said in the locality that some developers were (wisely or unwisely from a development point of view) interested in the kind of proposals embodied in Property and Estates Management's boards, but they were put off any active involvement when they discovered the Dicksons' connection. The facts of this matter must remain obscure. The Property and Estates Management advertisement asserted that 'If you're in North Sea Oil you're better off in Peterhead' and offered a brochure entitled *The Promise of Peterhead*.

Although the brochure was as vague as it was glossy, one fact was clear: Site Preparations had a development plan for the Peterhead area. It looks as if the company had evolved a private development strategy for the district on a scale that might be thought more appropriate to a public agency, or at least to an authority subject to some measure of public control. Their land-holdings would, if planning permission was given for specific developments, enable them to pre-empt most of the policy alternatives that might have been open to the County Council and later the District Council. Site Livestock Ltd (a subsidiary of Site Preparations) was, according to the *Aberdeen Press and Journal*[5] planning 'a project which will encompass large scale slaughter and storage facilities, canning, fertilizer and animal food manufacture, tannery and other by-product processing'. It was intended that meat products should be exported to the EEC and to Scandinavia by using a roll-on roll-off ferry with a terminal on the south side of the Harbour of Refuge.

Such plans were attractive because they promised alternative employment to oil and construction. Site Preparations had done their homework and discovered the shortage of meat killing facilities in the region and the need for new export outlets. The old Aberdeen County Council gave planning permission, in principle, for the abattoir and for hotels and housing, but detailed plans were not submitted for approval. The original 'plans' were maps with stickers affixed, and marked *hotel*, etc.

At the beginning of 1975 Site Preparations took the Secretary of State to court, accusing him of stealing their plans for the Harbour of Refuge and land nearby. The Secretary of State claimed that these plans had been sent unsolicited. These legal wrangles are of no consequence at this point of the analysis, except as an indication of conflicts to come.

More litigation was to follow: in 1974 'about a dozen' court actions had appeared in the Court of Session calling list by the end of September.[6] By June 1976 a number of cases concerning Peterhead were being heard. On 31 May Site Preparations sought interim interdicts to stop United Dominion Trust Ltd (who had taken over old Broad Street Securities) and their estate agents selling land in the Peterhead district. UDT said Site Preparations owed them £1.5 million upon which the monthly interest was £20,000 − Site Preparations had, in other words, defaulted upon payment of a standard security due to be paid by 1976. Site Preparations replied by saying that they held the title to the land and that UDT's attempts to sell it were 'extremely prejudicial' to the company which was currently involved in liquidation proceedings in the Court of Session.[7]

This particular dispute continued and in September 1976 UDT sought a warrant to sell Newmills of Sandford, Whitehill, Meethill and Damhead. Similar court action was outstanding between the parties in Kilmarnock, Alloa and Haddington Sheriff Courts and the Court of Session. On 5 November 1976 as a result of the Court hearing in Edinburgh, it was said at a public enquiry into the compulsory purchase, by the NSHEB,* of some of Site Preparations land, that the company was 'hopelessly insolvent'.

If industrial development had been a possibility in late 1975 or 1976 it may well have been held up by the litigation in which Site Preparations was involved − even an eager developer would have been unable to buy out the interest of United Dominion Trust because Site Preparations disputed it. The default of an ambitious and litigious private developer would have held back possible industrial or other developments in the Peterhead area, had such developments been contemplated.

The Regional Council encountered a further difficulty when in 1977 it was found that United Dominion Trust owned land that had been sold to the County Council by Site Preparation in 1974. The Chief Executive reported the mistake (failure to check title to the land) as 'inherited' from the old County Council, but, as the responsible official in the old county, it was very much his bequest. The ratepayers had to pay £278,025 twice over for a piece of land upon which to build a short road.

The land deal which attracted most local attention (partly as a result of a TV programme), involved an important local body, the Harbour

*The NSHEB rarely needs to use the CPO procedure to acquire land.

Trustees. This was in connection with the Arunta base at Keith Inch. The peppercorn rent of £1500 has been widely criticised in the town, but defenders of the negotiation point to the fact that the North of Scotland Hydro-Electric Board was paying £400 an acre to *purchase* land at the time. The main beneficiaries of the Arunta deal were the Harbour Trustees, who collected about £200,000 in dues in 1976 and, indirectly, the old Town Council, which received £160,000 in repayment of the interest on loans made to the Harbour Trustees.

Of course a substantial part of these dues were collected from the fishing harbour, but never the less the Trustees' deal with Arunta was not quite as the public imagined. That it was a good deal is attested to by Arunta's attempting quite early on to change the terms of the lease. The harbour developments are dealt with in more detail below.

In November 1971 Arunta International wrote a plan in execrable English for the development of the Bay Harbour: the first two phases of the plan entailed development at Keith Inch but the third entailed the development of the Crown land at the base of the south breakwater (where Site Preparations had planned their ferry terminal). Only the developments at Keith Inch went ahead. The Harbour Trustees were not entirely happy with Arunta and wanted to consider other applications. Arunta wanted to make an early start and could not raise funds until they had some kind of promise of a site. The negotiation proceeded from the end of 1970 until July 1971, and the lease was executed at the end of February 1972, when developments began at Keith Inch.

One other roadside sign hints at delayed or unsuccessful development or speculation: to the north of the West Road there stood a sign reading: 'A Buckwell Investments Ltd. Development . . . in association with Charterhouse Associates Ltd. (Aberdeen)'. Buckwell Investments Ltd was a subsidiary of Buckwell Holdings Ltd and, as its accountants had noted, entirely dependent upon the parent company, having a deficit of £428,553 and short-term loans of £1,000,256. Charterhouse Associates (Aberdeen), founded in 1973 with a share capital of £½ million is ultimately owned by the Charterhouse Group Ltd, a London-based enterprise with a capital of £35 million. The hint that Buckwell may have failed with this particular investment is provided by the emptiness of the field in which the board stands, and the 'For Sale' notice attached to the board.

The company was quietly realising its assets while it could, having decided in late 1974 that there was no money in speculative warehouse building or property dealing in the area. One of the directors, however,

through another firm, sold 43.9 acres of land to Aberdeen County
Council. He had purchased this land in 1973 for £210,000 and sold it
to the ACC for £450,000.

Leaving Peterhead to the northward, one passes the commodious
bungalows, built in the fishing boom of the early 1970s, on the right-
hand side of the road. To the left is a small area zoned for industry and
initially reserved for local non-oil related firms, which Arunta Ltd sought
to buy in May 1972 as part of the development already begun at Keith
Inch. Beyond this and a small corridor reserved as public open space is a
steep knoll overlooking the River Ugie, with a fine sea view on a clear
day, but utterly exposed to the wind on a foul day. Some new large
houses stand on this knoll behind boards announcing the site as a Rush
and Tomkins (Homes) Development. The land upon which they stand
is part of 8.01 acres purchased by the Peterhead Development Company
on 16 July 1976 for £84,552. Peterhead Development Company was
formed in 1974 and had a share capital of £10,000. The Rush and Tomkins Group held 2,500 of the B shares and the Scottish Northern Investment Trust 2,498 of the A shares. The two remaining A shares were
held by advocates of 6 Union Row, which is the registered office of the
company and the office of the solicitors for the Scottish Northern
Investment Trust. The contractor who built the houses at this site,
called Waterside, had gone bankrupt, owing about £9,000 to one local
building contractor and plant hirer.

To the south of Peterhead and adjoining the land belonging to Arunta
Properties (Scotland) Ltd is the estate of Longside, formerly the property of the Ellerman Trust. This land now belongs to Ronald Titcome,
the founder of the Arunta base. His intention was to farm the land and
reclaim some of the unused lands within the estate and he also hoped to
lease land for quarrying – a matter which has caused considerable local
controversy, because the lessee went ahead with quarrying, having withdrawn his (locally contested) planning application to do so. In every
direction around Peterhead therefore, land and its use has been an issue,
both in the sense of arousing popular comment and controversy and in
posing problems for planning.

More work needs to be done on land deals in Peterhead. Ownership
and land transactions are a matter of public record and the pattern of
land-ownership around Peterhead can be established with further research, but options on land are not publicly revealed and no systematic
study of this is possible without interviewing every property owner in
the district. Enquiries on such matters are notoriously likely to produce

false or misleading information or no information at all. Rumour is rife; stories of carpet-bagging agents with rolls of banknotes trying to persuade local landowners to grant options are common. One local garage owner was persuaded to grant an unlimited option upon his property to Site Preparations, so that when he wished to retire some years later he could not dispose of his property. Another researcher reliably reports that a certain public official was offering college students from Peterhead fees of up to £100 for information that they could glean from amongst their friends at home about possibilities for obtaining options.

At the end of 1977 it might seem as if the speculators had simply burnt their fingers in Peterhead. UDT, for example, held land which at its peak value was worth £10,000 an acre. The sale of this land today would not even cover the interest charges on the original loan to Site Preparations. New developers are moving in more cautiously: there is some hope of the development of offshore maintenance services in Peterhead which would entail some growth in warehousing, storage and light engineering in the zoned land. Such possibilities have been signalled by BOC's announcement that it intends to build a £2 million base extension to the south of the town. It is still possible that Scanitro will build an ammonia plant and the firm is keeping its planning application alive.

There are none the less visible consequences of zoning and speculation. Almost inevitably standards of agricultural management have declined in the zoned areas. The most dramatic example is the Buckwell field, which would probably have taken five years to get into good agricultural order had it not been purchased recently as a site for a supermarket. But a similar, if less dramatic, decline can be seen in areas adjacent to the zoned land. This suggests that owners or tenants see little future in their land or that they have already granted options on it and hope to see these taken up.

Concluding discussion

The observation was made at the beginning of this discussion that control of land passed out of local hands and that the new controllers held sufficient land almost to constitute themselves virtually a development authority. These points would be of little moment if the outcome was industrial, commercial and housing development. Jobs, houses and a growing economy may well outweigh considerations of local democratic control — especially when that control had produced very little in the past.

Little development ensued from the early phase of activity. This may be attributed to the failure of other developments to take place — most notably the ammonia and NGL (National Gas Liquid) plants, in the case of Peterhead, the refinery in Cromarty Firth. These developments depended on economic and political considerations outside local (perhaps outside national) control.

It might also be the case that in, for example, Peterhead, no development was intended by Site Preparations. The company obtained land and promoted the idea of development perhaps only with a view to selling at the increased values created by the idea. We have certainly located other putative developments which were intended for profitable sale after a limited input of plant and services. These limited developments were financed by institutions looking for a high return on loans.

No one hesitated to use the word 'speculation' when we spoke to property developers, although it had a variety of connotations. The policy of OIL was described by the director of another firm as one of buying land extensively in the knowledge that most of the purchases would entail a loss, but a few sales would bring in a multi-million pound 'killing'. They succeeded with the sale to BP in Shetland and Scanitro in Peterhead. Defenders of the property companies pointed to the risks involved; the companies incurred heavy debts with no certainty of a return. Local landowners did well; whenever they sold they made more money than they could ever have made before. If they waited hoping to make even more money by selling directly to developers they might have made nothing at all in the end.

Whilst this defence is partly valid and many companies had to sell land in order to meet their debts, it is not entirely so. Peterhead and Fraserburgh Estates, for example, were not taking much risk in buying Barclay's land when they were simultaneously negotiating with a buyer.

In arguing loss of local control and the growth of what were, in effect, private development agencies, we ignored the role of the local authorities. In Shetland the County Council took control entirely out of the hands of developers. In the Cromarty Firth the local authority upheld its right to plan by resisting development schemes that deviated substantially from their own intentions.

In Peterhead too the County and Regional Councils were seen to be in control. What happened behind the scenes is more obscure: the zoning of land may have been developer-led. The rapid growth of the

Regional and District Planning Departments would probably have ensured that detailed plans from Site Preparations would have been rigor-
ously dealt with — had they been submitted. The fact remains that the
town plan was evolved with development proposals of a grandiose, but
none the less vague kind in the planning pipeline. Once permission has
been given in principle it can only be withheld on technical planning
grounds. Therefore permission given in principle must influence the development of plans for an area.

The extent of private holdings is also remarkable in Peterhead. Site
Preparations and associated companies held nearly half of all land zoned
for industrial and commercial development and over half if one excludes
the publicly owned Dales Estate at the northwest corner of the zoned
land. With Arunta Properties (Scotland) Ltd they completely boxed in
the town to the south and southwest, as may be seen in Map 3.

One major effect of the developers'/speculators' activities was that
local authorities had to pay a high price for land. Shetland Island Council paid £2.3 million for land which a few months previously they could
have bought for £200,000 or less. In Peterhead the Dales Estate was
purchased for £525,000 when in 1972 it was on the market for £25,000.
43 acres of land for local authority housing in Peterhead cost Banff and
Buchan £450,000 in 1975 when in 1973 the property developer paid
£210,000. Of course not all the difference in price is accounted for
simply by speculation because a local authority must pay the market
price for land according to its use, and land for industrial development
is more expensive than agricultural land. None the less the market value
of land for any use is influenced by speculation. In a sense a speculator
must always succeed because his purchases push up local land values
and start price rises which increase the value of his earlier acquisition.
But whatever their impact on the local landowners and land market the
developers have done well from the public purse. If this money has not
enabled the companies to pay off debts or interest then their ventures
in the north and northeast have failed.

Local authorities which hold high-priced land without any development upon it have also failed. But they can service their loans from rate
funds, thus, until development does take place on the land the public
will continue to pay.

Finally, we note that the property developers buy local knowledge
and skills and try to give their ventures a local dimension, even if only
in nomenclature. In the Cromarty Firth OIL (in the form of CFD) made
a major *coup* in employing two officers of the HIDB. In Peterhead the

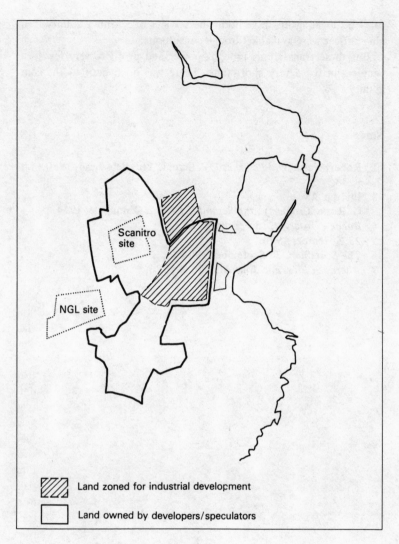

Map 3   Use and control of land, Peterhead

local law firm was able to supply the necessary knowledge although in the case of Site Preparations the aid was soon withdrawn (in fact, Site Preparations never paid the law firm for its services). The Scanitro deal, however, was highly profitable and may have resulted in some benefits to PFE's local agent. In addition the services of a former County Planning Officer were available through consultants.

But these transactions brought power and profit to very few local people and for a very short time. There was no benefit to the community.

Notes

1 Robert Neish, *Old Peterhead*, P. Scrogie Ltd, Peterhead, 1950, p. 38.
2 Ibid., p. 48.
3 G. Rosie, *Cromarty: the Scramble for Oil*, Canongate, 1974.
4 *Buchan Observer*, 13 February 1973.
5 23 November 1974.
6 *The Scotsman*, 30 September 1974.
7 *Aberdeen Press and Journal*, 1 June 1976.

2 Planners' dilemma

Introduction

Peterhead's geographical location was the reason for the oil industry's interest in the town and the industry's interest was the reason for the speculators' activities. But now that the land had been acquired, what was to be done with it? This was a problem both for the speculator and for the planner who hoped not only to exercise some degree of control over the speculator's enterprise but positively to *plan* the developments. By planning we mean anticipating and making provision for population change by providing houses, schools, roads, water, recreation facilities and so on. But in order to plan one needs data: how many jobs will be created, how will this affect the population, what will be the impact on local industry, how long will the developments last? These are but a few of the salient questions to which it was not possible to give accurate answers in 1973/4. In fact no one was quite clear about the planning questions to which answers were needed.

The planning needs of Peterhead depended almost entirely upon the *level* of petro-chemical developments. 'Almost' because the town would also be influenced by the harbour developments at the exploration, development and production phases of offshore work, and fishing remained a source of uncertainty. Given its importance we will concentrate upon the gas landfalls in discussing the uncertainty that was to be the byword for planning in the region in the 1970s. We start this chapter, however, by examining the first direct and public impacts of development in Peterhead.

Strathbeg

Before August 1972 the impact of oil in Peterhead was hardly significant. There were developments under way as the Arunta base was being
built and this would soon make its own impact. But the first issue to
stir the public into action was the proposal to build a gas terminal at
Crimond. The gas was to be piped ashore north of Rattray Head and
pass through the sand dunes between Loch Strathbeg and the sea to a
terminal on the old naval air station (HMS *Mergansa*) at Crimond. The
Loch of Strathbeg is said to have been formed in a single night when a
violent storm threw up the dunes and turned a piece of the coast into a
lake 3km long. The loch is an attractive feature in an otherwise unat-
tractive rural landscape. The dunes, however, are of scientific interest
and the loch itself has become an important location for bird life, and
especially for migrating birds.

The immediate local response to this proposal came from the *Buchan
Observer*, a paper that will hear no good of oil, in contrast to the *Aberdeen Press and Journal* (a Thompson newspaper) which will hear little
ill. The *Observer* headed an article on 15 August 'Rape of Peterhead'.
The article dealt with escalating land prices and asserted that 'There is
one hell of a lot of gullibility floating around the north-east just now,
aided and abetted by public figures, government, local and national'
and it went on to contrast the qualities of the Peterhead character with
that of the slick and ruthless operators who were moving into the area.

On 9 December the *Observer* was suggesting in response to the draft
structure plan that:

> All that Peterhead has gained from the Forties oil fortune to date is
> a boom in real estate and a rocketing cost of living . . . . We are in
> the throes of real-life Monopoly. New lairds are springing up like
> dragons' teeth.

By December, however, new interests had been drawn into the debate
about the Crimond site. Professor Dunnett of Aberdeen University called
for a careful consideration of the proposal on conservation grounds. At
about the same time news leaked out that the Ministry of Defence
intended to build a 'radio farm' on the Crimond site. But it was the
conservationist lobby that took the limelight; 27,000 signatures were
handed into the County Council by the North East Environmental Group
headed by four university professors. Amongst the 128 objections

received largely with reference to potential pollution hazards and the scale of the undertaking, was one from a keen ornithologist and owner of half the loch's shore. The County Council Planning Committee confessed themselves to be 'taken aback' by the size of the development, but the owner of the proposed 500-acre site — a county councillor — kept his own counsel and refused to comment to the press upon his own attitude.

At the beginning of February 1973 the British Gas Corporation held an open meeting at Crimond. Various objections were stated; the RSPB and the Nature Conservancy were especially concerned about the habitat of the rare Whooper Swan, but then other objectors mentioned rising prices (especially rents and rates) as an undesirable consequence of development. At about this time local opposition became organised with the formation of the Buchan Action Group under the chairmanship of the Episcopalian minister. According to the *Press and Journal* of 8 February, the aims of the BAG were:

> To oppose terminal or other projects which would adversely affect the loch of Strathbeg.
> To insist on delaying planning permission until a strategic plan of industrial development in the north-east has been published.
> To act as watchdogs on future developments.

In retrospect it is difficult to judge the extent to which this opposition represented local opinion. There were some dissenting voices suggesting that conservationists had biased the whole discussion and that the public were really in favour of the development. The research findings suggest that the public had been surprised and gratified to hear that they had such an important natural site upon their doorstep. The BAG , for example, found the Buchan Trades Council supporting them at the Crimond open meeting. They had expected the Trades Council to be pro-development because of the employment offered, and therefore anti-conservationist. But the Trades Council, which consistently supported oil-related developments, represented a view that has been expressed by many working people. Why ruin such an important site and spoil the habitat of thousands of birds when equally usable land was so close by?

The British Gas Corporation revised its plan in April and suggested moving to a site about 3½ km to the south of the Crimond site, just north of the hamlet of St Fergus. A public meeting in St Fergus gave the proposal 'an overwhelming thumbs up response' according to the

Press and Journal, 3 May 1973. On 17 April the *Press and Journal* reported that the Crimond plan was finally dropped and the Ministry of Defence had the final say, needing the Crimond site in 'the interests of national security'. The MOD's intentions had been known for some time and it seems that their interest was used by the developers as a face-saving way of avoiding a public enquiry. The councillor who owned the site did not lose by this decision: he had originally bought land from the Ministry of Defence, purchasing a total of 760 acres for £12,840 in 1970. The Ministry had been trying to buy the land back, but he thought they had offered him an inadequate price, especially as the land had been re-zoned for industry in the meantime (*Sunday Times*, 8 January 1978). He made a conditional sale to the BGC and this forced the MOD to make him a very much better offer than before. In 1973 he received £300,000 for 490 acres and used the money to purchase additional farm land.

Throughout developments such as these the *Buchan Observer* kept up its hectoring commentary of woe. In May 1973 'Peterhead is now an island in a sea of sharks' and on 5 February 1974, shortly after the *Press and Journal* had reported 'Set for Boom at St Fergus — Wee Village prepares for Golden Future', the *Observer* described Peterhead as a 'town living on its nerve ends in the politics of despair'. By 4 June, when the building of an ammonia plant was being discussed, the Secretary of the Conservation Society was calling for 'decent, honest prosperity and not worship of a "Golden Idol" ' (*Buchan Observer*). This the editor of the *Observer* echoed in declaring a week later 'Mammon is at large, utterly ruthless and subtle . . . . Euphemistically, it is called development.'

Comments such as these heightened suspicion of the developers and it was widely rumoured that BGC had chosen the Loch Strathbeg site in order to exhaust opposition before applying for the St Fergus site, which was their real objective. To anyone familiar with the activities of multi-national corporations such a Machiavellian scheme seems entirely plausible. It was certainly the case that no objections to the St Fergus proposal were received by September 1973, when the proposal had been in for five months. But the exhaustion theory is probably quite untrue. The Crimond site offered many advantages; the air station had large areas of hard-standing and services — sewerage, water, electricity — even if at a very low level of provision. The St Fergus site had none of these advantages, it was a field site with no services. Crucially the lower-lying St Fergus site created drainage problems and necessitated the piling of the foundations of every structure erected. Estimates of the cost of piling to the bedrock varied from £5 to £10 million per site.

In March 1974 the RSPB announced that the Loch of Strathbeg was to become a bird sanctuary.

St Fergus

The removal of the site to St Fergus did not leave Crimond deserted. The old air station was used as a pipe store and as a site for construction camps. It was reckoned that the St Fergus site would employ about 2,000 men at the peak of construction activity. Camps for 1,000 of these workers were provided by the three main contractors. Living conditions in these camps and the effects the presence of the construction population had upon the local population will be discussed in a later chapter. This accommodation was provided at the prompting of the County Council who foresaw considerable strain upon the housing resources of the area if no special provision was made. The worst outcome would have been unauthorised caravan sites which might rapidly have developed into insanitary slums of the worst kind, lacking all the physical and social amenities thought necessary by the local authority.

In the event the BGC allowed work to go ahead on the Crimond site and on a caravan site without planning permission having been obtained. The Planning Committee said that:

> Should such a contravention occur in the future, the Committee will not hesitate to take immediate enforcement action . . . . Although the committee appreciate the economic advantages in an early start to the project, they can not safeguard for the local community.
> (*Aberdeen Press and Journal*, 8 September 1974)

The local authority drew up specifications for the camps which were not, in the event, published until the camps were completed. The County Council recognised a further need for temporary housing of a kind suitable for families. This was needed to cater for long-stay professional and executive employees who would bring their families to St Fergus for the duration of the project. In addition local people might move to the area in search of work and need accommodation until — unlike the executives who might buy a house — they qualified for a council house. Thus the County Council built 150 mobile homes, similar to holiday camp chalets, on the road between Crimond and St Fergus. This — the Keyhead Estate — became something of a problem because, as part of the county (and later district) housing stock, it offered opportunities to house 'problem' people whose stay was to be far from temporary.

The developments at St Fergus have had wide implications for Peterhead and to understand these we need an understanding of what was being built at St Fergus and for what purpose.

The Anglo-Norwegian gas from the Frigg field is brought ashore by pipes on the seabed via a number of pumping and boosting platforms. The offshore work was carried out by the French Elf-Aquitaine companies (the loss of one of their platforms delayed the commencement of gas flow to St Fergus). On shore the raw gas is received in a terminal plant owned by Total Oil Marine at a rate estimated at 1400 million cubic feet per day. Here chemical slush consisting of butane, propane, ethane and natural gasoline is removed from the gas. The Frigg gas is dry, being about 98 per cent methane by volume so these 'natural gas liquids' may be removed from the site by road tanker to be converted into feedstock in the chemical industry. The gas is then supplied clean, dry and cooled to the adjacent British Gas Corporation site. Here the gas is further cooled, compressed (and therefore reheated), again cooled and then pumped into the gas grid when it becomes the 'North Sea Gas' burnt in kitchens and used as an industrial fuel.

Unlike the Frigg gas the Brent gas is 'wet', being high in the ethane, propane and butane and only 85 per cent methane by volume. The condensates from this gas flowing at about 600 million cubic feet per day cannot simply be taken away by road tanker. The implications of this will be discussed in the next section when we turn to the problem of planning for petro-chemical development.

Planning

In 1962 there were only two qualified planners in Aberdeen County Council and by 1972 this had grown to a department of about five qualified planners and twenty staff. From 1972 to 1975 the qualified planning staff grew rapidly to 35, but it is important to note what happened in this period: most planning was concerned with small-scale housing developments and occasional small industrial projects. In 1972 the staff available to cope with this at county level was smaller than the present Banff and Buchan District Planning Department. This small staff with conventional experience had to cope with large-scale developments in oil and gas plus the associated developments in commerce, servicing, transport and housing. Simultaneously they had to contend with the reorganisation of local government in May 1975. Peterhead Town Council was never a planning authority and if it had been all the proposals would have fallen outside its jurisdiction except the Keith Inch base development. Prior to reorganisation the County Council was responsible for planning in Peterhead, and afterwards it was the Banff and Buchan District Council and Grampian Regional Council.

The County Council and its regional and district successors had to cope with two kinds of planning problem. Firstly, they needed to process specific applications like the St Fergus gas terminal. Here scientific and engineering knowledge of a kind not normally available in a local authority were needed to test the assumptions and proposals in the application. Secondly, they had to make their own plans to deal with the consequences of the proposed developments.

Local authority planning departments are not staffed with petrochemical engineers and they are, therefore, largely dependent upon consultants and the industry itself for some of the technical information that would enable them to make decisions. This was certainly the case in granting planning permission for the St Fergus development. The application was scrutinised by the County Council but neither checked by independent consultants nor subject to the vetting and probing of a public enquiry. The county may have failed to appreciate the scale of the development and the extent to which the proposals embodied untried technical innovations. The failure to carry out a full environmental impact study may have been responsible for the failure to examine the effects of high-power radio transmissions from Crimond, which became a safety question in mid 1977. The omission of a full environmental impact study was partly compensated for by Shell, Total and BGC agreeing to pay for a report by Cremer and Warner which was submitted to Banff and Buchan District Council in May 1977, *The Environmental Impact of the Natural Gas Terminal at St Fergus*. This report helped prepare the district for the later Shell development at St Fergus and was especially important in informing the planners of the kind of control that would be needed over possible future developments close to the terminal. But the Shell/NGL pipeline application raised problems of environmental hazard which could not be dealt with locally. Thus the Regional Council objected to the proposal when it was submitted to the Department of Energy which was dealing with the application under the Pipelines Act. This ensured that a safety and hazard study was carried out by the Health and Safety Executive.

The Scanitro enquiry (Chapter 3) was conducted with a relative lack of information and few special reports were commissioned independently of the applicants' own submission. Banff and Buchan District Council had to draw on an enquiry into noise for all their technical evidence. In the case of the NGL plant application the applicants' own impact study was felt to be inadequate on questions of advanced technology and again Cremer and Warner were asked to report. This was paid for partly by Shell, who also carried out a study of the harbour and met the full cost of the consultant's assessment of the study on behalf of the local authorities.

For more general planning the region and district had to rely upon its own resources. 'Uncertainty' might be taken as the keyword for the county, the region, the district and for the town of Peterhead from 1970 onwards. The problem of uncertainty derived directly from the question of the level of petro-chemical development. By 'level' we simply mean that at each stage of landing and processing oil and gas there is a choice to be made between taking the products away for use elsewhere, or processing them at least one more stage on the spot. We can understand some of the implications of this by reference to Figure 1: let us first follow the major product of the gas fields, methane, through its possible stages of processing.

As we saw above, methane is 'North Sea Gas' and when the NGLs have been removed it may either be fed into the gas grid to be burnt as a fuel or it may be converted into methanol and ammonia. Methanol is the raw material for solvents and resins, ammonia the basis of fertilizers and acrylics. The choice of levels in Peterhead are therefore (1) NGL separation plant and gas pipeline to south, (2) NGL separation plant plus ammonia/methanol plant, (3) separation, ammonia/methanol plant and further downstream production of fertilizers, etc. Choice 2 does not exclude methane going into the gas grid and some being converted— the choice is not exclusive. For each level additional construction work is required to produce various discrete plants which would need a permanent workforce. The same is true of NGL processing, although the choices are technically more complicated.

(1) The NGLs, once separated from the methane, may simply have their natural gasolene removed and then be piped away to the south. (2) They may, alternatively, be converted into Substitute Natural Gas

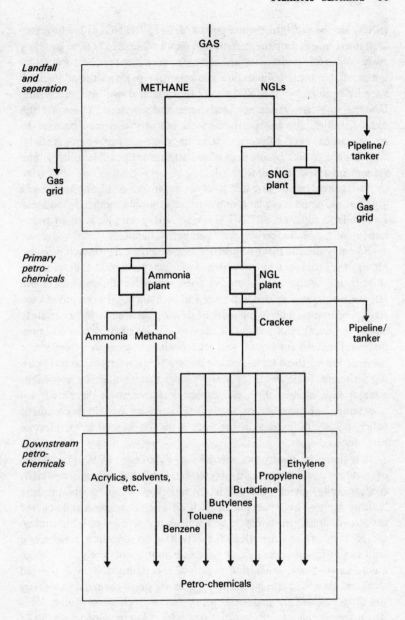

Figure 1   Stages in gas recovery and processing

(SNG) and be fed into the gas grid as fuel. (3) The NGLs may be separated into ethane, propane, butane and naptha. These may then be taken away by tanker under pressure and/or at very low temperatures (they are gases at normal temperature and pressure). (4) With the addition of an ethylene cracker the NGLs may be broken down into propylene, benzene, toluene, butadiene, butylenes and ethylene. These are the basic materials for the petro-chemical industry and may be used to make plastics, dyes, solvents, detergents, synthetic rubber and artificial fibres — the choice becomes so wide that it is not possible to list all the subsequent stages in their use.

At its simplest the possible levels at Peterhead were, for methane; a pipeline and ammonia plant, downstream ammonia/methanol products. For NGLs; a pipeline, an SNG plant, an NGL plant plus harbour terminals, an ethylene cracker, downstream petro-chemicals.

Not only did the local authorities lack experience with such developments; they did not know which developments would take place or what their consequences would be. There was no handbook which would tell them the number of temporary and permanent jobs created at each level. Uncertainty about the level of development leads to uncertainty about the number of jobs to be created, the eventual size of the population, the need for houses and schools. It is not simply a question of more or less of these social resources, because at a certain level of housing provision, for example, a new sewerage system might be needed, or a major expansion of the water supply. In other words there are quantum jumps in provisions, or, in planners' language, certain thresholds at which a smooth expansion becomes a sudden leap in infrastructural provision and costs.

It is therefore hardly any wonder that Peterhead saw the production of a number of reports and projections (and that others were made confidentially) giving widely different estimates of jobs, population and housing needs. The reports dealt with different time-spans and covered various administrative areas and so we can not give a simple chronological account of the projections for Peterhead to show how these varied with the different estimates, guesses and rumours of levels of development. Landfall and separation plant was a certainty for the Peterhead area, and an SNG plant a possibility. Primary petro-chemical processing was proposed and planned for. Downstream petro-chemical production— which would change the character of Peterhead — remained a matter of uncertainty and rumour.

Let us look, then, at the various plans and projections which tried to make sense of the future in conditions of uncertainty, starting with the last 'pre-oil' scheme for the Peterhead area the *Deer District Plan* published in April 1972. It proposed to encourage development away from the coastal towns by making Mintlaw (13 km west of Peterhead) a rural development location, in a triangle with Fraserburgh and Peterhead. The *Plan* was geared to the needs of a declining district and assumed that 1,200 jobs would be lost from agriculture and that 1,850 new jobs would be needed. This set a target of 3,050 jobs in the decade 1971–81. The plan was entirely overtaken by events. Let us summarise this and subsequent plans with particular reference to projections for jobs and housing, in Table 1.

TABLE 1

Report	Period covered	Estimate of jobs	Estimate for housing
Deer District Plan,	1971–81	1,200 lost 1,850 needed for target of 3,050	
Buchan: the Next Decade, April 1973	1976–81	6,000 new jobs	
Buchan: the Next Decade, Modified Recommendations, December 1973	1974–81	1,500–5,000 new jobs 3,500 best estimate	3,110 houses
		Also proposed expansion of land zoned for industry.	
Peterhead Expansion, Proposals 1973		See *Buchan: the Next Decade*	(i) 580 by 1976 plus 480 (?) (ii) 1,900 in phase 2
Buchan Impact Study, June 1975	1975–82	Level (i)* 1,900 Level (ii) 2,600 *see text below	To be revised downwards to (i) 2,880 houses (iv) 3,470
Regional Report, June 1976	1976–81	2,500 increase	3,400 for Banff and Buchan as a whole.
Housing Needs, November 1976	1976–81	1,500–2,200	2,271 houses

In June 1975 there appeared the most comprehensive evaluation of the possibilities for Buchan; this was the Economist Intelligence Unit's *Buchan Impact Study* commissioned by Aberdeenshire County Council

and the Scottish Office. The impact study hypothesised five different
levels of development in Buchan:

    (i) Existing developments — resident oil companies, supply bases,
       terminals at St Fergus and Cruden Bay, Boddam power station.
   (ii) Above plus the Scanitro Ammonia Plant and expansion at St
      Fergus.
  (iii) A Shell-Esso NGL separation plant in addition to (i) and (ii).
  (iv) Added a second NGL plant and a methanol or second ammonia
     plant to (iii).
   (v) The highest level of development adds an ethylene cracker com-
     plex to (iv).

Level (v) constitutes the level at which petro-chemicals might 'take off'
in Buchan, the products of the ethylene cracker complex opening a
multiplicity of opportunities for downstream developments, and Peterhead would become recognisable as a major petro-chemical centre.

The EIU, however, reckoned the cracker complex to be of low prob-
ability. It was hard to disagree with this because Buchan does not have
easy access to markets for petro-chemical products and the level of
development implied might have been thought, as a matter of national
policy, to be more appropriately located in the Scottish midlands or
elsewhere. The *Impact Study* observed that 'none of the land presently
zoned for industry is suitable for locating the specialised [petro-chemical]
plants' on grounds of safety and disturbance. Thus the Town Plan would
have to change.

According to the *Impact Study*, the existing Buchan plans provided
housing in excess of likely demand by about 1,500-2,000 dwellings.
Aberdeen County Council intended to complete 4,424 houses on un-
constrained sites (sites without problems of ownership, planning or
servicing), 3,350 of these in the years 1976-8. On *all* sites the total was
7,473 (5,259 in 1976-8). The only circumstance in which the two
highest levels of development would exceed the 1976-8 completions
was if 80 per cent of the housing demand was in the public sector, in
which case private houses would be left empty if the Buchan forecast of
completions was met, with 53 per cent of the provision in the public
sector. The *Impact Study* (p. 44) therefore recommended downward
revision of housing targets after 1976. But the intra-district distribution
of dwellings is also of considerable importance because it is plainly less
likely that housing need can be adequately met if houses are not located
where they are needed.

If all the constraints on building could be overcome, then the town of Peterhead alone, the study argued, could cope with the whole expansion at any level of development. The most pessimistic estimates, however, would put Peterhead under strain, but the housing need uncatered for in Peterhead could be met in Mintlaw (eight miles west of Peterhead) (pp. 106–7). Looking ahead to 1982 and taking into account housing built before 1976 the *Impact Study* concluded that Peterhead's then current programme alone would cater for most of the Buchan requirements. Such an outcome would vitiate an important part of the intentions of the policy outlined in *Buchan: the Next Decade, Modified Recommendations*, namely to spread some of the housing expansion into the rural areas, thus providing for limited growth away from the dominant coastal towns.

In October 1976 the Regional Council printed a confidential document which examined the implications of twelve possible levels of development. This examination was prefaced by a note on the employment implications of specific activities: about 50 jobs are created by an NGL plant; 120 by an ammonia or methanol plant; ammonia or methanol developments downstream create 300 jobs and a cracker 250. The cracker, as mentioned above, represented a possible employment take-off with between 400 and 1,000 jobs and perhaps a further 900 jobs as a result of the multiplier effect.

The twelve levels of development ranged from 'no plants' to (twelfth) NGL, cracker and downstream, ammonia and downstream and/or methanol and downstream methanol developments. For the eighth level of development Peterhead harbour becomes inadequate to cope.

When the possible effects of the twelve alternatives had been listed the confidential document discussed the variations in supply of condensates: if it increased to 9 million tons in the late 1980s by a gas-gathering pipeline landfall (bringing in the gas that would otherwise be flared off at sea) in addition to the Frigg and Brent gas, there was a possibility of two crackers, a second NGL plant and many other downstream developments. But a new harbour would be required to be built before this stage could be reached.

This latter comment on the need for a new harbour underlined the need for careful, long-term planning. The covering letter with the document and para 1.1 of the main text alluded to this and a plea made in the Regional Report that

The Secretary of State is requested to concentrate planning effort at a national level on the development of an oil and gas strategy for

Scotland, in conjunction with the appropriate planning authorities and to indicate the role of the Buchan Area in that strategy to enable the Regional and appropriate district council to amend their programmes accordingly.

This national strategy has not yet been developed. In retrospect much of the above discussion seems to have been abortive; the North Sea oil bubble in the northeast appears to have burst. Not only is the NGL plant likely to be built in Fife but the Scanitro plant (if built at all) seems likely to be situated in Denmark. None the less, in 1978, in the light of uncertainties about the Mossmorran development, the question of pipeline and other hazards and problems of world markets, a new round of planning for petro-chemicals in the Peterhead area was under way — at least the possibilities were being considered anew, on paper.

Concluding discussion

The uncertainty about level of development and the problems this creates for planning is not only important for regional and district planners but for the sociological model one adopts of the events taking place in Buchan.

Gas with more or less condensates (from the Brent and Frigg fields respectively) arrives at the terminal (St Fergus). It is then separated into methane and condensates. The methane can be burnt as 'natural gas' through the national gas grid — in which case St Fergus has merely been the transhipment site for the removal of a basic or primary product. The condensates may be simply stored and then tankered out (as for the present 'dry' Frigg gas) or piped away for separation into natural gas liquids. Again, in either case, St Fergus tranships raw materials. None of what has been described in this paragraph would constitute 'manufacture', 'processing' or 'downstream development', the natural products are only processed to the point where they are most conveniently transported.

Therefore, without the NGL plant, the locality is without the chance of self-sustained growth based on petro-chemicals. Without the ammonia plant it loses even the consolation prize and becomes an area entirely dependent upon, and subordinated to, economic developments elsewhere. In fact the ammonia plant would have been more than just a 'consolation prize' but a finite development with the possibility of further limited and finite developments very much more in scale with the employment needs of Peterhead than an open-ended petro-chemical development based on NGLs.

This, then, is the background to events in Peterhead. Regional and district planning authorities creating their planning machinery and attempting to plan in a state of chronic uncertainty, with little help from central government and lacking either experience or expertise in the field of petro-chemicals. Between 1972 and 1976 there was on paper an expansion of developments, population and infrastructure and then a decline of all these. By 1977 the future was still unclear with some developments proposed (pipelines), others in abeyance (ammonia plant) and yet others undecided (a gas-gathering scheme to collect all the gas from the smaller fields). Over and above this there is perhaps the feeling that there are other possible developments as yet unpublished, but perhaps equally uncertain.

3 The community mobilises

The decision to build the largest gas terminal in Europe a few miles to the north of Peterhead did not excite much local interest. The proposal to build an ammonia plant on the southwest boundary of the town did. In October 1973 the local MP confirmed that rumours of an ammonia plant were well-founded. The *Buchan Observer* (9 October 1973) greeted this with a headline typical of its response to proposed oil-related developments 'It's now hush hush ammonia: Killer gas for Howe o' Boddam'. A year later no site for the plant had been decided upon. It seems that the consortium which wished to build the plant was having difficulty in acquiring land because most of the development land was, as we have seen, in the hands of property companies, who were asking prices which the company was unwilling to pay. One result of this delay was that Superfos, the Danish partner, withdrew from the project (*Aberdeen Press and Journal*, 26 October 1976). In November the consortium Scanitro obtained the 102 acres of Wellington Place for £400,000 from Peterhead and Fraserburgh Estates.

On 19 November 1974 a press conference was organised at the offices of Peterhead Town Council by the Town Clerk. The conference was to tell the press about the Scanitro undertaking. The Episcopalian minister, a town councillor, was also acting as a stringer for the *Buchan Observer* and gained admittance as a press man.

The consortium consisted of Norsk Hydro (Norway), Supra (Sweden) and Scottish Agricultural Industries (ICI), who had an option to participate. The plant would cost £50 million and would produce 350,000 tons of ammonia for export to Scandinavia, although SAI would retain 50,000 tons if it exercised its option (which it did not). The ammonia was to be exported through Peterhead harbour. The construction of the plant was expected to take three years and to need 400–600 men who would live in a camp on the site. About 100 permanent jobs would be created.

The convener of Aberdeen County Council, who was also chairman of the North East Scotland Development Association (NEDSA), said after the meeting that he welcomed the project 'with open arms' because it was the type of permanent development which the area needed.

Some local people felt, however, that it was not the kind of development they needed. At the beginning of June a chemical plant at Flixborough in Lincolnshire was the scene of a violent explosion followed by an almost uncontrollable fire. Fifty people were feared dead and 2,000 people evacuated from the village, which suffered extensive damage from the blast of the explosion. The number of deaths was actually 29 but a further 40 people were injured. This explosion received headline coverage in the press and press reporting and discussion continued as the enquiries into the disaster progressed into the autumn of 1974. The Episcopalian minister had lived near Flixborough and clearly remembered the unequivocal and emphatic assurances given by the developers there that there would not be the slightest risk of pollution nor the danger of any accident. The Revd Alexander was therefore quite properly suspicious of the bland assurances given at the press conference and he introduced the only contentious note into the conference by asking if Scanitro knew that they were proposing to build their plant within three-quarters of a mile of a housing development.

The Flixborough disaster provided a focus for local opposition to the proposal and this opposition found a willing ally in the *Buchan Observer*, which in September had called for a moratorium on all developments other than oil and gas. An Ammonia Plant Action Group was formed under the chairmanship of Dr Manson, a local GP. There was a meeting in the Peterhead North School on 11 December organised by the Buchan Conservation Society at which it was agreed that a petition should be raised amongst the electors of the town. Accordingly at the end of December a petition was launched objecting to the proximity of Scanitro to the town. Over 5,000 signatures were collected; four out of every five approached signed the petition and three-quarters of the 10,000 electorate were contacted.

Dr Manson himself did not object to the Scanitro development as such. However, he felt it to be too close to the town, visually intrusive and on such a scale as to dwarf all other aspects of the town. He was also concerned about the short-term demand for construction labour created by the plant, because he feared this might permanently damage local industry.

Other objections were not so clearly formulated and there was relatively little data from which potential objectors could derive their arguments. It therefore fell to a small group of activists to marshal information in a form that could be used to lodge an objection. The initial reaction to the development of not trusting the company to create another Flixborough would clearly be less than adequate when it came to a public enquiry in which objectors would be subject to cross-examination.

The objectors received much help from the factory inspectorate and their technical advice underlined the need to avoid simple emotional appeals to the hazards of disastrous explosions. The *local* expert was Michael Smith whose family had run the woollen mill for 150 years. He was a conservationist who had informed himself on many aspects of the world fuel situation and was building himself an energy-conserving house.

There was also generally felt anxiety that an ammonia plant was the thin end of the wedge and that other, more extensive, petro-chemical developments would be bound to follow it. This fear was aggravated in February 1975 when at a public meeting in Peterhead the County Council refused to say whether further developments were planned for the Peterhead area or not. The Town Council 'expressed their concern' about the development on the grounds of siting, pollution and the lack of infrastructures. There had been delays in providing essential community services, a sewerage scheme had been cut on the grounds of its cost and the local authority was already committed to heavy expenditure to service existing developments. One councillor later suggested that Scanitro should establish a fund to meet some of these costs, as a condition of planning permission. George Baird, a Labour councillor, whilst in favour of any development that created jobs, felt unable to commit himself because of the lack of information. This lack of information plainly troubled the whole council and fed the suspicions of many townsfolk.

The Buchan Trades Council supported the proposal because it was committed to the expansion of employment opportunities and saw Scanitro as the beginning of a change in the occupational complexion of the area, and the level of wages.

Opposition to Scanitro opened a number of conflicts in the town;

the manner in which names were collected for the petition was criticised on the grounds that 'people will sign anything' and that shopkeepers presented the petition to customers in a way that made it hard to refuse to sign. It was also felt by some that the petition against the *siting* of the plant drew much support from people who were emotionally opposed to the whole scheme as a result of the publicity given the Flixborough explosion. The Revd Alexander was also quite unpopular because of what was alleged to be an ecclesiastically proprietorial attitude towards a town marked by deep religious differences; so too was Bob Johnston who was a leading light in the Buchan Conservation Society and who was felt by Peterheadians to be a bit of a crank and, anyhow, an outsider who, living 5 miles away, had no business interfering in Peterhead's affairs. In fact Johnston, a vet, was mainly troubled by the loss of agricultural land and the transformation of a rural society into an industrial society. Some of the sharpest opposition to the APAG and the deepest hostility to its activities was expressed by people who themselves were opposed to Scanitro or ambiguously disposed towards it. This proved to be only the first example of the deep personal hostilities which divided Peterhead. It is important to bear in mind that Peterhead was and is a very small town in which activists all knew one another. Interpersonal relationships are therefore important in conflicts over large issues. The Deputy Chairman of the Harbour Trustees, for example, needed considerable persuasion to give evidence to the enquiry, not because he doubted the Trustees' interests in the Scanitro proposal but because he did not want to get 'on the wrong side' of people. The Action Group needed to vote on various questions when in theory they had only come together because they were agreed in their opposition to Scanitro. One prominent member dropped out when it seemed that the proposal would be approved in spite of objections. Others hesitated to support the group because they distrusted certain leading members. Few of the committee were prepared to make the necessary efforts of research and lobbying that were needed to make the campaign successful.

In mid-January 1975 members of Aberdeen County Council, Peterhead Town Council and Buchan Trades Council flew to Norway to see the ammonia plant at Porsgrunn, operated by Norsk Hydro. Mr Johnston refused to go because he felt it would be said that his opposition to the scheme had brought him benefits in the form of a free foreign trip. He observed that everyone who went to Porsgrunn came back saying how nice it was — a reference to at least one councillor who dropped her objections and instead supported the development.

At the beginning of March 1975 the county gave permission for the development subject to thirteen conditions. In May the new Banff and Buchan District Council gave permission in principle although they expressed misgivings at only having Scanitro's own information on which to base their decisions. In June Grampian Region's planning committee approved in principle. On 17 June the public enquiry began. An enquiry was required because the planning authority wished the development to go ahead despite the existing plans for the area. A direction was therefore required from the Secretary of State to permit this.

The Scanitro enquiry, June 1975

The Director of Planning for Banff and Buchan began by saying he did not think that the proposal would make an unbearable impact on the district. Existing plans anticipated a population increase of 4,280; only if this increase equalled 4,500 would there be a significant impact on the existing planning problems (Report of Proceedings, pp. 499 ff.). There was plenty of land zoned for industrial development and the consequential housing developments.

The Factory Inspectorate asserted that there was little risk of fire or explosion, although a small explosion within the plant might be possible. Ammonia is difficult to ignite or explode. It is, however, an irritant and the main hazard was seen as spillage at the loading jetty. For this reason Hogie Oystese, Norsk Hydro's expert on chemical transportation, suggested the need for close co-operation between users and harbour authorities.

The harbour was to become a major focus of the enquiry as it was in the subsequent NGL application. Captain Shepherd, who had been asked to report on the harbour by the SEPD, pointed out that the two offshore supply bases would have to restrict mooring and movements to allow both ammonia and NGL tankers to turn. (The first of the Shell–Esso NGL applications had already begun and was later withdrawn but at this time NGL tankers were becoming part of the discussion.) About seven tanker movements would take place every week, disrupting fishing movements, but not extensively. Michael Smith interjected at this point in the enquiry to ask what would happen if one fishing vessel slipped in before the others were stopped and thus enjoyed special advantages in the market. One could calculate that such a consideration might arouse anxieties amongst the fishermen to whom access to the market is vital.

For Scanitro it was said that no further downstream developments were to take place. Ammonia would be produced at 1,150 tonnes per day and shipped away to Scandinavia. The highly competitive nature of ammonia products leads to the siting of plant near the feedstock. The production of ammonia itself is very capital intensive, but 100 people would be employed by Scanitro, of whom about 70 would be recruited locally. According to Professor A. T. McIndoe (a consultant to Scanitro) the area was already designated for industrial development and therefore committed to the provision of infrastructure; the increased rateable values would lead on to loss of rate support grant, but it none the less made local government less dependent upon central government. Peterhead, according to McIndoe, was the biggest industrial site in Scotland except for Hunterston.

The Action Group shared the services of the distiller's QC, but it was Michael Smith who led the most effective opposition in the enquiry. The Scanitro development was not the 'utilisation of an indigenous resource of the UK continental shelf' but imported Norwegian gas. None of the product would be delivered to the UK. Yet the firm would receive a minimum of £10 million in grants plus further discretionary grants of up to 50 per cent (although Smith doubted if these latter grants would be available for capital-intensive development). There would be heavy public expenditure plus the use of 7 megawatts of electricity subsidised by the British taxpayer. The great advantage of Peterhead to local businessmen was the stability and quality of the local labour force. Local firms had made heavy investments in labour, for example it cost £1,000 to train a worker at one local firm. Scanitro would create job-destruction; one firm had lost 84 workers since January and had only been able to find 25 replacements.

Smith scored further points off Scanitro's witnesses by making the very important observation that three witnesses speaking on the harbour did not know Peterhead's tidal range (the difference between highest and lowest tide, easily found in Admiralty Tide Tables). He also alluded to the power of multi-national corporations and their collusion with government.

However good this case and however well-prosecuted it was unlikely to impinge very deeply on the enquiry because it related to general questions of policy rather than the technical details of the application. At the most it would be taken as a cogent expression of local opinion. Smith presented a theory of the 'developments' proposed that fits well with sociological theories of underdevelopment: a powerful foreign firm merely uses a weak nation as a site for its activities, exploiting the government and local labour, damaging the local economy in the pursuit of profit to be made on products geared for a foreign market. Smith argued, however, not as a sociologist nor as an opponent of capitalism but as a small capitalist facing 'unfair' competition from a multi-national in collusion with the state. He felt as threatened by the power of the trade unions as by the multi-nationals.

The Buchan Conservation Society — who had been unable to obtain any advice from their national organisation — objected to the use of good agricultural land for industrial development (as did the Department of Agriculture and Fisheries) and feared further encroachment upon land for the servicing of the present project and for further associated developments.

Two witnesses spoke of the effects of the development on the life of the community. Dr Taylor, a local general practitioner, alluded to increased drinking amongst young people and a 'retreat from community' by professional and business men who were increasingly commuting into the town rather than living in it. He also underlined the lack of investment in public services. The reply to this latter point was that development would generate more rates — but this is a spurious point because increased rate income leads to loss of rate support grant as Professor McIndoe had said.

The Revd J. F. Miller spoke for Deer Presbytery. He introduced into the enquiry a concept of 'culture shock' (Report, p. 2218):

> a new culture overwhelms an environment and its inhabitants too quickly . . . so quickly that the new social possibilities and cultural realities are operative before they can be grasped, appreciated, corrected and evaluated. This communal cultural shock shows its evidence by manifestations of helplessness, or of despair, or at least of resigned defeat.

Miller then went on to suggest that this was now the case in Peterhead. Furthermore (p. 2220): 'industrial development would mean the introduction through temporary labour forces of a new criminal element into an otherwise law-abiding community.' It was all more than Peterhead could cope with, the Boddam Power Station, St Fergus, the increased activities in the harbour and with all this went a shortage of housing that forced many young couples to start married life with their in-laws.

The contributions by Taylor and Miller are important from the point of this book; are the changes they report actually true or coming true and, if so, can they be attributed to oil? Claims similar to theirs have a very general currency and the special 'social problems' created by rapid development have exercised local, regional and national authorities. Data on these are of considerable interest to policy-makers, planners and the general public and we will give special attention to them in chapter 6.

There was one moment of light relief — or high drama — in the enquiry when a representative of Site Preparations intervened to say that the whole enquiry was invalid because they owned part of the land upon which the Scanitro pipeline to the harbour was proposed and had not been given notice of the enquiry.

The Reporter in his report to the Secretary of State said he thought the local opposition to the scheme as represented by the signatures on the petition was largely based on ignorance and misapprehension because the petition was raised before details of the project were known. He did, however, recognise the considerable feeling of local resentment at the large expenditure proposed when cuts were being made in public services. He found the Buchan Impact Study superficial and he was not 'greatly assisted' by the Scottish Economic Planning Department. On the question of siting and all technical matters relating to safety, noise, pollution, etc., he felt either completely satisfied or that the balance of interests lay in Scanitro's favour. He duly recommended that subject to specific planning conditions the application be approved. His very last words were to be a source of considerable conflict, he recommended 'That steps be taken forthwith to institute a single harbour authority responsible for the control of the whole Peterhead Harbours.'

The Secretary of State duly gave an Article 10 direction which allowed the District Council to extend its land zoned for industry to include the Scanitro site. Banff and Buchan in collaboration with the Regional Council acted upon the Reporter's observations upon costs. In March 1976 they refused Scanitro permission to proceed with the plant until 'adequate contributions to the town's infrastructure' were worked out (*Aberdeen Press and Journal*, 11 March 1976), and in July 1976 presented their terms to Scanitro for the relative shares of infrastructural costs between the developers and the Councils.

To date Scanitro have shown no signs of actually building their plant in Peterhead. The world market for ammonia would not justify the development, although the consortium has kept the permission 'alive'.

By 1977 Norway was feeling the effects of general recession and in the General Election campaign of that year it seemed as if there would be considerable pressure on future governments to insist on landing and processing gas from the Norwegian sector in Norway.

Perhaps we may draw a few preliminary points from the discussion so far:

1 The town of Peterhead as such had no standing in the discussion of the Scanitro development. The development was outside the boundary of the burgh, which was not a planning authority anyhow. After local government reorganisation, the town ceased to exist as a separate administrative authority. The people of Peterhead therefore had no formal *local* channel of representation at the enquiry; they depended either upon consultation or voluntary action.

2 The townspeople were at a considerable disadvantage in lacking both detailed information about the ammonia plant and expert understanding of it.

3 The major interests involved were foreign and their considerations relative to world markets rather than local interests.

4 The 'threat' of an ammonia plant was not sufficient to overcome antipathies within the community and thus no unanimous voice could be heard.

5 The scale of development proposed and the sums of money involved were of an order that could not readily be grasped by inhabitants of a small town.

All these factors point towards the relative lack of control, and, perhaps as important, feeling of lack of control over events in Peterhead. The ammonia plant was something happening to them, it was to bring few tangible long-term benefits that they could see, and yet change their town and their lives in a way that they did not understand. Even for those in favour of the development, it was not something over which they could have much influence – they were not even required to speak at the enquiry. Much as they might want the jobs and the prosperity the project promised they could not influence its coming, which depended upon economic and political factors well beyond their control.

A larger and more contentious planning application had started on its way towards a public enquiry before the Scanitro question was settled. But no equivalent to APAG was organised to contest it. After the Scanitro enquiry the general feeling was that there was no point in resisting development proposals – they would go through whatever the locals said. And yet the Scanitro enquiry could have been mere 'warming up', a training session for a determined anti-development group. But political energies seemed spent and conflicts within the community had come to the surface; we will see more of these later. A few individuals only went on to fight the next application.

Shell/Esso

Shell/Esso originally applied to the District Council for permission to build an NGL plant and ethylene cracker. This was then withdrawn and on 13 February 1976 a new application was submitted for an NGL plant alone. At this time no final decision had been given by the Secretary of State on the Scanitro application.

Shell/Esso said they wished to build a £93 million NGL plant at North Collielaw, but it was clear from press comments that they had not entirely given up the idea of ethylene manufacture also. Furthermore, Mr W. C. Thomson, a director of Shell International Chemical Co. and chairman of Shell Chemicals UK Ltd, read a paper in April 1976 which spoke of the suitability of Peterhead for the location of an ethylene plant. On 1 March Shell's plans were presented to the people of Peterhead in a meeting arranged by the Regional and District Councils with the help of Shell. Shell/Esso argued that Peterhead was a suitable site for development because (i) it was a development area, (ii) it had a suitable harbour and (iii) the Boddam power station was a safety valve through which NGL products could be burnt off. By this time the Secretary of State had already called the application in for decision on the grounds that the application was of national interest with broad policy implications. This course of action did, however, deprive the local authorities of a chance to process the application and formulate a considered response. The *Press and Journal* described the meeting as 'somewhat fiery' because a former councillor called the proposal a disaster for Peterhead. He went on to suggest that Peterhead harbour was totally unsuitable for the shipping of liquid gases by large tankers. He claimed to have spoken to some 500 people before attending the meeting and to represent their views. Another questioner wanted to know how much money the applicants would be receiving from the taxpayer and was told that the Secretary of State might make a grant equal to 20 per cent of the total cost.

Ex-councillor Forman received some support for his view. Cremer and Warner reported to Grampian Regional Council in mid-March and said that the harbour facilities were 'barely adequate'. The harbour serviced the fishing fleet, two bases, the Hydro Board, possibly an ammonia plant and now possibly an NGL plant. An ethylene cracker in addition raised such serious questions of harbour capacity that an alternative site for the cracker might have to be considered. The consultants found the proposed NGL plant acceptable in terms of siting, pollution and noise at North Collielaw, but they made recommendations concerning bunding of storage tanks, the need for a back-up to the proposed safety system on the loading arrangements and for a full firefighting service. They also judged the harbour to offer potential hazards for large vessels in bad weather.

The Secretary of State ordered the public enquiry in April. The public response was less than that to Scanitro. A reporter from the *Press and Journal* found the people 'punch-drunk'. The feeling of powerlessness was well-founded. Objectors in the Cromarty Firth had successfully opposed a refinery in Nigg but the Secretary of State had overturned his Reporter's recommendation, and the Secretary of State had similarly given approval for a runway extension at Edinburgh airport despite successful local opposition. In Peterhead it was felt that he would give approval to Shell/Esso whatever was said at the enquiry. This feeling was reinforced by knowledge that Shell already had a 1,000 million dollar contract approved by the government to supply NGLs as chemical feedstock to Omaha, Nebraska.

The Regional Council welcomed the proposal but the District Council were against it unless they could be satisfied on certain points. Firstly, they wanted the plant sited on a disused airfield 5 km west of Peterhead. This would provide land for further developments from the NGL plant, such as an ethylene cracker and downstream activity. It would also prevent the NGL plant blocking alternative industrial developments to the southwest of the town. Secondly, the district required clarification of the harbour hazards and the treatment of spills: thirdly, reassurance that no serious job-destruction would occur locally, and finally they were not satisfied with certain aspects of the initial impact of the proposed plant. The Harbour Trustees shared the cost of Counsel with the District Council and were thus joint principal objectors. They noted that they had not been consulted and that they wanted no harbour works that would risk closure of the fishing harbour entrance. Any accident resulting in such a closure should entitle fishermen to compensation.

There were corporate objections also: the Aberdeen Conservation Society feared that gasoline vapour from loading in the harbour would result in the closure of Crosse & Blackwell — this fear was echoed by the firm who pointed to the risk entailed for 500 jobs in the town. The distillery feared loss of its water supply for distilling and cooling.

The Scottish Economic Planning Department wrote a letter in which they said that priority would be given to fishing traffic in the use of the harbour. One organisation in support of Shell was Buchan Trades Council which welcomed the development as 'creating employment and generating a new vital base for the future economy of the North East of Scotland'. They did, however, ask for the elimination of hazards and for the developers to pay a larger share of infrastructural costs. A submission in support came from Councillor Baird. His letter, none the less, raised a number of questions concerning firefighting, the use of labour and the provisions for incoming workers. He also noted that the Secretary of State refused the town funds to meet even its present infrastructural needs.

Michael Smith wrote in strong terms that there was collusion between commercial interests and the Scottish Office against the wishes of the local community. The sale of NGLs to Nebraska was already contracted, so Shell clearly believed permission to be a foregone conclusion. The calling in of the proposal before the planning authority, the District Council, could process it removed local democratic rights and bred 'local cynicism, disgust, apathy and a feeling of total impotence to the vital matters affecting our future safety and stability'. Amongst specific effects Smith cited the destruction of local jobs.

The sense of impotence was expressed in eight of the twenty-one private submissions on behalf of twenty-eight individuals:

'As the citizens of Peterhead now say "what good is objecting, the decision has already been taken and our so-called regional officials will do what they are told by Whitehall and the Scottish Office".'

'(the environment) appears to be of little or no consequence to big business which can steam roller its way over the wishes of a large proportion of the population . . .'

'we are informed we have got to have an Ammonia plant, whether we like it or not, a feeling of apathy, brought on by the apparent inevitability of such projects, is making itself felt.'

The private objectors raised objections on the grounds shown in Table 2.

TABLE 2

Harbour	Hazards	10
	Disruption of use	5
Environment	Pollution, etc.	10
	Danger	6
Agriculture	Loss of land and disruption of agriculture	6
Other	Effect on local economy	6
	Change in nature of community	2

The public enquiry began on 25 May 1976 in Peterhead. Shell's Counsel summarised the District Council's objections and went on to assert in a rather ambiguous way that 'this development is fully in the national interest and is not contrary to the interests of the people of Peterhead and district'. The early part of the enquiry was especially concerned with safety and the use of refrigerated pipeline from the plant to the jetty. Shell resisted the district's request for them to move to the airfield because they did not wish to use a very long cryogenic pipeline. From the discussion at this stage of the enquiry it seemed as if the development was near the limits of engineering experience and some slight but reasonable doubts might remain about the hazards of the pipeline and jetty. This *seemed* to be the case because Shell's witnesses developed a habit of delegating answers to questions to later witnesses — a practice which eventually irritated an otherwise very patient Reporter. A number of Shell witnesses appeared for one day only and a reading of the transcript of the enquiry suggests that there had not been very close liaison between witnesses and certainly no conference before the enquiry itself.

Part of the evidence, however, related to labour demand and is of particular interest, although the figures given by Shell were not entirely consistent. The size of the construction force would depend on the extent to which the plant could be fabricated off-site and then assembled as modules. These modules would be brought to Peterhead on a single-carriage road deemed quite unsuitable for the then level of traffic in 1937. During the year Shell's estimates had risen from 600 to 1,100 to 1,400 and these figures excluded work on the jetty which was the responsibility of the SEPD (Report, pp. 1025 ff.). Seventy per cent of the peak labour force would be electricians, riggers/erectors, scaffolders, insulators and painters. Welders were likely to be in short supply and Shell were worried by their effect on the local labour market. They hoped to recruit from Edinburgh, Glasgow and northeast England; if it was legal they thought they might restrict the employment of locals to protect the local economy and perhaps give preference to returned migrants. Two-thirds of the workforce were expected to live in a camp at North Collielaw (pp. 950-64). Shell did not wish their men to share a camp with any others, as this, they felt, was a recipe for industrial unrest (p. 1097). Because travelling men work for the money only and therefore seek maximum overtime Shell thought it might be difficult to restrict Sunday working, but if such work was restricted they would need a larger workforce (p. 998).

There would be construction work at both St Fergus and the proposed plant, so the peak labour force would comprise (p. 1101):

NGL	1100 - 1400
St Fergus	750 - 950
	1850 - 1950 (sic)

The plant when operational would employ about 77 people; 15 management and administration, 40 operators and supervisors, 22 maintenance. At the St Fergus gas terminal there would be 43, 8 management, 20 operators and 15 maintenance. About 60 of this total permanent workforce might be employed locally and the number might increase if and when local men were promoted.

William Bell, managing director of Shell UK Ltd, responded to local fears by expressing the personal view that no downstream developments based on NGLs would take place in Peterhead because a nucleus for a Scottish petro-chemical industry already existed elsewhere — he felt that not even an ethylene cracker was likely (pp. 57-8).

At an early stage Michael Smith interposed some very hostile questions in order to make the point that Shell had disregarded planning requirements at Anglesey (by failing to bury a submerged pipeline). Furthermore Shell's international status enabled it to avoid statutory price control by exporting and reimporting products. The company was, in his opinion, 'above Governments and above Enquiries' (pp. 172 ff.).

It was on day 10 of the enquiry that Shell's case began to disintegrate: it appeared that the Harbour Trustees had not been consulted over the use of the harbour by tankers, although each tanker movement would disrupt the harbours by about one hour. Shell claimed to have consulted the 'harbour authority' but by this they meant the SEPD who, whilst owning the Harbour of Refuge, were not the pilotage authority nor the owners of the fishing harbour. Also in Shell's original submission on the harbour it seemed that they were satisfied with the existing provisions; but a single copy of a further plan submitted on the first day of the enquiry suggested that harbour reconstruction was necessary. Captain Dickson of Shell International Marine Ltd was cross-examined on the original submission (p. 1851):

'Was it known that the harbour was going to have to be reconstructed for your purposes in November 1975' — 'it was not absolutely concluded at that time that the harbour would have to be reconstructed.' . . .

'when do you say it first became apparent that the harbour was going to have to be reconstructed?' — 'When we had consultation in Holland on, I think, the 11th of March of this year, and we had the first appreciation of the work done at Delft.'

The proposed alteration to the harbour entailed raising the height of the south breakwater to prevent overtopping, plus rockfill with concrete cube armouring on both sides to a distance of 30 metres. In addition there was to be a spur projecting northwest into the harbour from the end of the south breakwater. The cost was estimated at £47 million. No tests had been done on these alterations to see how they affected other harbour users and the Wallingford model upon which the first tests were done did not include the fishing harbours. This all entailed a quite fundamental change in the nature of the enquiry, which in theory was concerned with the use of North Collielaw Farm. The alterations would be of interest to the NSHEB, Scanitro, and the BOC and ASCO bases, also the Harbour Trustees (as representing the fishing harbour interests and as the pilotage authority) and the RNLI whose lifeboat ramp pointed directly at the proposed spur. None of these had been consulted on the new proposal and nothing could be said of its likely effects on them until the tank tests were completed at Delft University.

The enquiry was adjourned so that the tank tests could be carried out and interested parties consulted. Shell/Esso also agreed to pay for independent consultants to assess their tests. On 11 November Shell announced the abandonment of their Peterhead plans in favour of a new site in Fife. In February 1977 the Reporter wrote to the Secretary

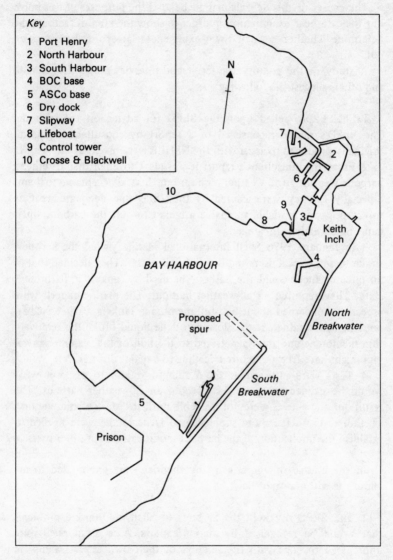

Map 4   Peterhead harbours

of State in terms highly critical of Shell who had, in his opinion, proceeded with the enquiry unreasonably, knowing that the harbour modifications were essential and yet not knowing whether test results were available and not consulting interested parties.

The causes of this debacle may lie beyond the purposes of this book but they do bear examination in order to show the kind of factors that determine whether a small town experiences larger developments or not.

A study of the enquiry transcript and interviews with participants and others suggest the following:

1 Shell-Expro relied upon the SEPD for advice on the harbour. The SEPD's material consisted of a report by consultant engineers, originally produced to assist with the NSHEB jetty proposal.

2 From the consultant's report it was clear to Shell that Peterhead harbour was not an all-weather harbour but this was acceptable to them, especially as they were assured by the SEPD that adequate weather forecasting was available to enable a tanker to quit the harbour when dangerous weather threatened.

3 In January 1976 Shell International Marine visited the harbour, spoke to some local users and the harbour master. They decided that in no circumstances could the harbour be used by tankers in its present state. They required an all-weather harbour. The marine experts were especially concerned about half-full membrane tankers leaving the harbour in bad weather. If the sloshing of the liquid broke the relatively thin membrane and then penetrated to the hull of the vessel the metal hull might have freeze-fractured, leading to a major disaster.

4 Tests were initiated at the Wallingford tank to discover what would be needed to make the harbour an all-weather harbour. The results of these tests were not available until the start of the enquiry. Further tests on the more sophisticated Delft model were needed to establish the implication of the harbour modifications for other users.

In the course of enquiries many theories were expounded to me about the withdrawal of Shell.

1 The SEPD over-sold the harbour to Shell, minimising problems that would be recognised by marine experts. As a result Shell were unprepared to cope with the findings of their own experts and were perhaps a little less likely to heed Cremer and Warner's comments.

All the remaining theories depend on the harbour modifications being used as an *excuse* to leave Peterhead.

2 Pressure was put upon Shell behind the scenes in order to move the project to an area of high unemployment in Fife, where the government needed votes. This version cannot be entirely discounted and the local MP believed that there was 'hatchet-work behind the scenes' (personal communication). No evidence can be found to support this contention however and there seems to be no reason for Shell to bow to such pressure in a way that brought discredit upon them. An open-policy decision could have been made by the Secretary of State.

3 Esso was interested mainly in an ethylene cracker and when they were told (off the record) that they would never get one in Peterhead they lost interest in the whole project, thus undermining Shell's initiative in the enquiry.

4 The District Council required Shell to sell their land to the Council and have it leased back; this is true. The theory continues . . . This was unacceptable to Shell who therefore found a way out. There is no evidence for this.

The prospective Labour candidate said that 'a great opportunity had been lost in the area which needed long-term stable jobs' and he attributed the loss to the negative attitudes of the Councils and conservationists (*Aberdeen Press and Journal*, 12 November 1976). He was correct but only in a perverse way: if no one had objected to the proposal and it had been accepted without a public enquiry, Shell would have started before the marine experts assessed the harbour, they would then have been committed financially and forced to carry out costly alterations — which may have delayed the start of that use of the harbour for an unacceptable period. There is no doubt that if they had gained planning permission speedily Shell would have gone ahead and been caught by their own ineptitude, but the unintended implication of Dr Bonney's comment is that none of the other questions of safety, pollution and economic consequences would have been discussed in public. The nature of the harbour alterations would have appeared too late for the other users to be able to stop the major undertaking at North Collielaw.

Ironically, Scanitro noted Shell's problems and made their own enquiries at Delft. They came up with plans at half the cost and with a much shorter lead time. Shell did not explore this possibility. The experts do not, however, always agree. The consultants who assessed the

Delft tests said that the Shell proposals would have subjected users to 'serious hazards, restrictions and delays' and would have adversely affected areas west of the north breakwater — at the entrance to the fishing harbour. The only safe solution he could see was building an additional breakwater to seaward (*Aberdeen Press and Journal*, 4 November 1977). Had Shell waited for this report they would have been faced with prohibitive costs and a very long lead time.

. The enquiry may accurately be called a fiasco; why was it so? Shell's joint undertaking with Esso probably created problems. The project manager knew the Peterhead site, but Shell-Expro was a new company lacking experience. It may have made less use than it might of the parent company's legal department's experience of planning applications. The marine branch at the Hague acted as consultants and their word was decisive on the harbour. The fact that Shell never pulled their case together in time suggests a lack of experience or undue haste. If Esso lost enthusiasm Shell staff may well have lost their commitment for the venture. This is surmise.

The lack of co-ordination *was* crucial. Had the marine experts visited Peterhead earlier, and spoken to experienced local users and the pilotage authority, the enquiry would either have been delayed or had better data before it, and the applicants would have been able to evaluate it before the enquiry. Once the mariners said they needed an all-weather harbour Shell was potentially saddled with extra expense, a longer lead time and many questions, and — in the event — questions that could not be answered in time for the enquiry. The best they could hope for was an adjournment followed by favourable reports on harbour modifications. But they cut their losses and moved to Fife.

In the sociological literature and more radical analyses the multinational corporation is usually presented as an efficient and ruthless monolith. The NGL enquiry suggests that, however powerful and ruthless they may be, one oil company was unable to organise itself to cope with a public enquiry. This inefficiency must have reassured Michael Smith, but it amused no one — not least the Reporter who awarded the objectors' costs in full against Shell.

There is no doubt that without the planning control available in the UK to scrutinise such developments the town of Peterhead might have found itself deprived of a main current source of prosperity in its all-weather fishing harbours. Fishermen of many nations too might have found themselves without refuge between the Tay and Cromarty Firth.

Whatever the arguments may be for or against the development of petro-chemical based industries in Peterhead the failue of the NGL plant to materialise ensured that no such industries could develop.

With the failure of Scanitro to build and of Shell to gain planning permission, the major bases upon which decisions relating to the provision of housing and other consequential planning decisions were based fell.

4 The Peterhead harbours

Peterhead harbours have a long history beginning with James VI's Charter to Robert Keith and George, the Fifth Earl Marischal, 'privilegium et libertatem unum portum erigendi' in 1587. Various Acts of Parliament laid down the constitution of the Trustees, the rights and duties of the Trustees and the charges they could levy upon vessels using the harbours. The jurisdiction of the Harbour Trustees extended over a substantial portion of Peterhead Bay which in 1873 (when the jurisdiction was most recently defined) was unenclosed water. Other Acts also gave the Trustees power to raise money for harbour improvements, although these were not always acted upon and fresh legislation was needed.

In the Peterhead Harbour of Refuge Act 1886 the Admiralty were empowered to use convict labour to enclose Peterhead Bay by breakwaters, thus creating a harbour of refuge. This changed the physical and pilotage conditions to which the Trustees' powers had been directed, but for as long as the Harbour of Refuge was unused except by occasional vessels seeking shelter from stress of storm (for which no dues are payable) this created no problems. The Trustees collected the dues from ships that broke cargo in the harbour, according to the terms of the 1873 Act. The spending beach, a flat, sandy beach at the south of the bay, became a recreational spot for local people, providing safe swimming in very cold water. The so-called 'lido' area was a popular playground for Peterheadians, a place for a walk or a picnic and somewhere to take the dog. The harbour itself provided good water for dinghy sailing.

The Prison Commissioners were, by the 1886 Act, empowered to erect a prison to provide the labour. This might be regarded as an early example of regional development policy. The Harbour of Refuge was completed long after the last sailing vessel might have used it in 1956, but it was used by the navy in war time and Admiralty moorings were maintained there. In 1960 the Admiralty offered the Harbour of Refuge to the Harbour Trustees, who under the 1886 Act already had rights to levy dues in the harbour for the duration of their indebtedness. The Trustees had debts and they had insufficient income to maintain the massive breakwaters and so they declined the offer. The Peterhead Harbour of Refuge (Transfer) Order, 1960, therefore transferred the harbour to the Secretary of State for Scotland and in so doing prepared fertile ground for conflict in the 1970s.

The prosperity or poverty of the people of Peterhead has always depended upon the harbours. According to Neish's history, *Old Peterhead*, there were major harbour improvements undertaken in 1739 and 1770. Acts of Parliament in the late nineteenth century further empowered the Trustees to develop the facilities. In the 1820s Peterhead was at its peak as a whaling port and the last of whaling activity is just within the memory of some of the oldest inhabitants today. Peterhead then became a herring port with 262 boats in 1836 and 480 in 1878. In 1920 there were 399 boats employing 1,377 fishermen and in 1930 187 boats and 963 fishermen. Alongside fishing and curing were the associated transport, cooperage and ice-making activities. There was a thriving Baltic trade in cured herring — local curers producing specific cures for specific Baltic towns. A herring canning factory was opened before the First World War and was purchased by Crosse & Blackwell in 1919.

The creation of jobs onshore is illustrated by Table 3. The figures do not include boats landing but not registered in Peterhead, which also contribute to the town's employment and prosperity:

TABLE 3   Fishing and employment

	No. of boats	Fishermen	Crosse & Blackwell	Total onshore
1940	119	366	479	1,100
1950	163	811	347	1,132
1960	123	700	720	NK
1970	106	496	800	1,070

In 1972 the inshore fishing fleet moved to Peterhead (see below), resulting in an immediate rise in landing of over 20 per cent. From 1971 to 1976 landings rose from 41.5 thousand tonnes to 61.8 thousand tonnes and in 1977 to 71.3 thousand tonnes.

Prosperity has been matched by poverty. In December 1931 the Fishery Officer reported to the Secretary of the Fishery Board for Scotland that: '50% of the fishermen are destitute and unable either to pay their bills or raise loans against their boats.' The banks would accept only houses as security for loans. The Fishery Officer also noted that the local fishermen were reluctant to apply either for government funds which were available to help them in their occupation or for public assistance to help them in their personal need.

In the 1950s and 1960s Peterhead was poor, with many young people leaving to seek work in the south. There was a notable exodus to Corby when Stewart and Lloyds opened a steel works there and locals still recall the shock of seeing train-loads of young people depart. The Harbour Trustees had to take substantial loans to maintain the harbour and improve its facilities. Some of their loans were backed by the Town Council on the grounds that the harbour and the town stood or fell together. The fishing industry was not the topic of my study, but it features in it, so it is perhaps worth commenting further that fishing has become more capital-intensive, requiring larger boats with more sophisticated equipment and that this in turn has led to over-fishing. The profitability of fishing now depends largely on rising fish prices, not rising catches. Secondly, Peterhead experienced deep division and personal trauma in religious crises in the 1960s when the community was divided by the teaching of Big Jim Taylor (later Archangel Taylor) amongst the Brethren — many of whom were fishermen. One result was that some Brethren, rather than be yoked with unbelievers by sharing the capital cost of larger boats, left fishing altogether. Furthermore the high costs brought the large fishing companies and trawler-owners into partnership with fishermen in Peterhead.

Whilst Peterhead, now a town of over 14,500 people, is dependent on fishing, it is not solely dependent on it, as we will see in Chapter 6. In 1973 42 per cent of the working population was in manufacturing, compared with a national average of 38 per cent.

A major change overtook Peterhead harbour when the inshore fishing fleet moved to the port from Aberdeen and started landing in 1970. The move from Aberdeen was based on conflict between the trawlers and the inshore fleet. The Aberdeen Fishing Vessels Owners Association operated a priority berthing scheme for their members and also charged 'strangers' higher dues for the use of dock facilities. In 1965 the Saturday market was abolished, and this affected the inshore boats because they normally made short fishing trips and returned for weekends. The trawlers stay out for weeks and can time their return to fit the markets more flexibly. Crucially the Aberdeen dock porters were unionised and controlled the off-loading of boats through a closed shop, and were able to make what the inshore fishers regarded as excessive charges for their services. As a result the fleet largely moved to Peterhead. With the passage of time and the healing of relations between inshore and deep-sea fishers who have had to organise their interests jointly *vis à vis* the EEC, the question of union porters has been increasingly singled out until it is now presented as the sole reason for the move to Peterhead.

The port may be entered 24 hours a day, 365 days a year and is never closed by weather. There are very few porters in Peterhead, so unloading is done by crews or their friends or relatives, and sometimes by schoolboys earning pocket money or by shift workers from the engineering works. A crewman who does not wish to wait to unload his boat may pay a substitute to do it for him. All payments are in cash and the labour entirely casual.

There were about 400 boats in 1977: about 30 from Denmark, a handful from Northern Ireland and the remainder Scottish boats. The harbours themselves have been altered; the North Harbour entrance has been closed and the wall between North and South Harbour removed; the fish market has been covered and a second expansion is planned. More boats want to land in Peterhead than the present market can accommodate. The Trustees hope to encourage more foreign boats to use the harbour in the future. Fish is now taken by lorry to the Paris fish market (arriving fresher than French landed fish) and as far as Turkey. This has resulted in a considerable expansion in haulage and haulage employment. The income of the harbours has risen to over £½ million per year and has enabled the Trustees to pay off some old debts. The income, however, depends upon the fleet remaining. The fleet's loyalty depends solely upon the availability of easy marketing and the lack of union control of portering. Thus any development which threatens either to close the harbour unpredictably or to unionise the port will result in the fleet moving away.

In 1972 the Harbour Development (Scotland) Act was passed through Parliament in four months. The Act only covers one and a half pages (unlike all the Peterhead Harbour Acts which are very long and wordy). In effect it grants the Secretary of State unlimited power to develop harbours 'made or maintained by him' and alter any other statutes (by repeal or amendment) as he may consider 'necessary or expedient'. A statutory instrument was published in 1973 as the Peterhead Bay

Harbour Development Order, which altered the 1886 Act to enable the Secretary of State to buy and sell land, to erect buildings, grant leases and to charge dues in the harbour. This was needed because the 1886 Act restricted leases to three years (presumably to ensure that only temporary buildings absolutely necessary for building the breakwater were erected).

This last provision was to be a potent source of conflict, because it meant that both the Trustees and the Secretary of State had the right to levy dues in the Harbour of Refuge. It seems that the draughtsman of the 1886 Act had not taken full account of the 1873 Act and that the 1972 Act overlooked this. Had the Secretary of State specifically sought to extinguish the rights embodied in Paragraph 81 of the 1873 Act, which lays down the limits of the Trustees' jurisdiction, he could, of course, have done so. There is no doubt also that the Trustees would have fought this in the courts. When I discussed this question with officials and employees of the SEPD, the opinion given was always that the Trustees' rights had been extinguished. But this would only be the case in the Harbour of Refuge. The Trustees' rights to levy dues on cargo crossing their land were not affected and they held thereby an important weapon in the conflict. The actual course and outcome of this conflict can only be understood in the context of the developments that took place in the harbours.

Offshore oil operations require substantial sea-borne support. In the exploration phase supplies are needed for exploration rigs, and in the development stage materials are needed for the building of platforms and the laying of pipes. A pipe-laying barge, for example, needs about six supply boats in support; pipes, bitumen, anchors, cables, diving and welding gases, food, general spares all have to be shipped out. Only about four day's supply of pipes and bitumen can be held on a lay barge and so continuous supply is necessary. Similarly with drilling and production platforms, a continuous supply of drilling mud and cement is needed in addition to food and general spares. When the oil fields are fully in production the bases will be needed to provide regular repair and maintenance services for offshore installations.

To sustain the necessary supplies it is necessary to have a base with deep-water jetties, warehousing and cranes. Because the offshore operators do not necessarily wish to engage in extensive onshore operations, they will expect the base to supply stevedorage, customs and excise clearance, to handle cargo and shipping dues and to co-ordinate supplies, hold stock and provide security services. Thus whilst an offshore operator may set up a field operating headquarters in a supply base he will not employ many hands: the operator's employees will be managerial and professional staff whose main focus of interest is the offshore activity. The base supplies a service 'package' to meet all their needs.

The potential for such a base in Peterhead was seen by Arunta International, Site Preparations and the Secretary of State. The latter believed that there might be a shortage of base facilities, and sought to establish a base in Peterhead 'in the national interest'.

In 1971 Arunta proposed a three-stage plan for the development of the Harbour of Refuge. The first phase involved building warehousing and office accommodation on Keith Inch, using the south pier of the fishing harbour (the South Harbour). The second phase entailed providing more substantial berths at dolphins on the inside of the north breakwater, capable of taking 12,000 ton vessels, the building of a helicopter pad, further storage and workshops and domestic facilities for vessels. The third phase included a new jetty at the south of Keith Inch and extensive developments at the base of the south breakwater.

At their March meeting in 1971 the Harbour Trustees turned down the Arunta proposal. The Trustees had already received tentative enquiries from larger and better-known firms about development on Keith Inch. These enquiries came to nothing and the Arunta proposal was the only definite application. The Trustees, however, were unsure of the financial standing of Arunta. Arunta tried to press the Harbour Trustees, stressing the need to make an early start on a base to serve the more northerly oil fields that were already opening up. The Trustees meanwhile made enquiries about the finances of Arunta. On 3 July 1971 there was a meeting between NESDA, the Harbour Trustees and Arunta at which the Trustees asked for plans to be in by 18 October to be considered alongside other applications. Arunta said that this delay would be too long. On 24 July it was agreed to go ahead and negotiate with Arunta and on 3 November it was agreed to give a 99-year lease to Arunta. The change of heart by the Trustees led to much local speculation and the *Buchan Observer* asked on whose side various named Trustees were. Originally Arunta had asked for a 5-year lease, but this proved insufficient when they sought financial backing. The 99-year lease, however, seemed to ensure support from merchant banks. The lease was terminable at six months' notice.

The rent for the Keith Inch site was agreed at £1,500 a year. This, many local people felt, was the Trustees selling their birthright. Throughout the period of our study we were told by Peterheadians how the

Trustees had been fooled into giving away Keith Inch to an eager developer. But the thirteenth clause of the lease also made all Arunta's cargo dues payable to the Trustees. Thus the Trustees' fortunes were tied to those of the base. After coming into service in 1973 in temporary accommodation and occupying their own purpose-build accommodation from March 1974, the base immediately began generating very high revenues. In 1973 3,000 coated pipes were to be landed — well in excess of expectations. In 1973 the Trustees were able to pay the Town Council £52,000 and in 1974 £26,087. Not all of the revenue that made these repayments possible was derived from the base; in 1973 oil accounted for less than 1 per cent of the Trustees' revenue, rising to a maximum of 30 per cent in 1976 and falling back to below 10 per cent in 1978. The suggestion that the Trustees should share in Arunta's profits originated from the Town Clerk (who was also Clerk to the Trustees) and the County Council (who had misgivings over Arunta's financial status).

Arunta was run as a non-union firm. According to Titcombe, the founder, one needed good communications in a firm and this was best achieved by adopting dictatorial attitudes. As well as believing in dictatorship, Titcombe believed that everyone should work together in partnership and there was no need for workers and management to divide into 'camps' — men with a grievance should elect a spokesman and thrash it out with him. He was therefore very anti-trade unions, which he saw as outside organisations interfering in the work of a firm. In his plans for the North Sea Terminus (with Peterhead and Fraserburgh Estates) he had listed one of the services to be provided amongst thirteen onshore support facilities '(9) Labour force (non union) . . . .'

There were certain advantages in working for Arunta. Titcombe was a paternalist dictator and provided overseas winter holidays for employees (and their wives) who had to work in the summer when the base was at its busiest.

Relations between Arunta and the Harbour Trustees were not always smooth. In May 1974 Arunta forbade traffic to use the north breakwater, but the Trustees maintained it was their right to control access. In the same month the Trustees noted that Arunta were reclaiming land without permission, to the hazard of existing harbour users because infill was being washed into the harbour. It was agreed that Arunta should be ordered to remove the infill. This does not appear to have been acted upon and in June the Trustees agreed to await a report before taking further action. In July they voted 3 to 4 against making

Arunta remove the spoil forthwith but agreed to an engineer making an inspection and recommendation at Arunta's expense. In February 1975 the engineer reported that the fill was washing out and causing instability but it was agreed to hold informal talks to find a solution.

At the end of 1973 both the Trustees and the Bay Management Company agreed to make a demand in strong form for Arunta to pay overdue charges. Issues such as these were clearly dividing the Trustees and some were thought by others to be in the pocket of Arunta. Another divisive issue was the amendment to clause 15 of Arunta's lease to give Arunta twenty years' security of tenure rather than the existing six months. This was in May 1972 when Arunta said that they could not raise more than £1 million with only six months' security. The Trustees were equally divided and the Chairman gave his casting vote in favour of Arunta's request.

Whilst these disputes continued, Arunta was seeking funds. One prospective partner was BOC (British Oxygen Company), who joined the venture in 1973. In the middle of November 1974 it was announced that the base was to be known as Peterhead (BOC) Base Ltd. Arunta had been bought out and, except for a small residual shareholding, the base had become entirely a BOC undertaking. The reasons for Arunta withdrawing are not easy to establish. It may have been that Arunta was in business purely to set up openings for other firms: thus by entering the market quickly in Peterhead it had been able to sell at a good profit to an operator who was equipped and financed for a development that Arunta could never have undertaken. Titcombe strongly denies this speculative interpretation of his venture and said that he was becoming disillusioned with the lack of vision shown by people in Peterhead. Furthermore he found BOC extremely difficult to work with. There does seem to have been a clash of management styles. Titcombe claims that BOC were excessively rule-bound and bureaucratic with little local autonomy. The BOC management deny this and it certainly is the case that the Base Manager has full operational autonomy — although he needs Board sanction for major investment decisions. Titcombe's free-wheeling benevolent dictatorship contrasts rather sharply with BOC's 'enlightened capitalism' and it may be over this that the main clash took place.

BOC claim not to be anti-union and encourage trade unions. At present they have only a works council in the base. The works council originated from a mass meeting to voice objections to the 1974/75 work schedules. A working group of six was set up which modified management's proposals to the men's satisfaction. The group of six had been so successful that it was decided to keep it going and the group became a works council. BOC recently asked the works council to reconsider their views on unionisation but was told by the council to back off. BOC management suggest that the works council feel that they have a strong voice and a control over their work situation which they would lose if official unions came in from 'outside'. My letters to the works council asking to meet members were never answered and thus I had no opportunity to hear the men's side of the BOC account. BOC also spend a lot of money on training, including business games to show the men the problems of the industry, instruction on grievance procedure, etc. Although they claim day-to-day control to be very autocratic, the BOC management also discuss their business plans with the works council and then sound out their opinions on matters of policy. This is not the Arunta style as described to us by Titcombe. Our best guess is that there was a clash of nineteenth- and twentieth-century capitalist methods between the partners and this led to the dissolution of what must have been a very uneasy marriage.

Arunta and Site Preparations had seen the opportunities for developing the harbour at about the same time and Site Preparations had proposed a scheme for developing the south side alongside the various industrial and commercial plans it had for the Damhead and Upperton sites. A letter signed by Jock Smith on behalf of Site Preparations was received by the Trustees on 8 April 1972 to this effect, but Smith was directed to the Secretary of State because the Trustees had no jurisdiction over the parts of the harbour in question. Titcombe had reckoned to amortise his development over a period of fifteen years with 1978–82 being the peak years. He came to the conclusion quite soon after setting up in Peterhead that he could not make a development on the south side pay in fifteen years so he dropped the idea. Meanwhile NESDA had identified a need for the expansion of offshore servicing and pressed the Secretary of State to take action. The County Council might have taken the opportunity to develop a base in the Harbour of Refuge, but as a conservative body it was fundamentally opposed to local authorities going into business. It was therefore left to the Secretary of State to take action. The Secretary of State's first action was to pass the Harbours Development (Scotland) Act of 1972 to enable him to develop the Harbour of Refuge — which from about this time became known as the Peterhead Bay Harbour. In June 1972 consultants were asked to evaluate proposals for the use of the south side of the bay; their report was published in August. In September a public meeting was held in Peterhead to explain the developments and in October negotiations were held with potential users of the harbours as a servicing base. In November an advisory committee was formed to assist the Secretary of State and a 15-year lease was agreed with the Aberdeen Service Company (ASCo) for the site on the south of the bay. In January 1973 work began on reclaiming 25 acres of ground at the south side, at the root of the breakwater and below the prison. In July 1974 the Labour Secretary of State opened the ASCo base on the site. ASCo is a subsidiary of Sidlaw Industries (Dundee). ASCo pay a rising rent to the Secretary of State who intends to recover his £3 million investment in fifteen years. Because of the high rent the cargo dues are remitted by the Secretary of State.

The ASCo development was conceived very much in the 'national interest' and, unlike the private proposals, it was unrelated to adjacent developments. It would not have been appropriate for the Secretary of State to become involved in regional and district planning, but he did establish the Advisory Committee for the Bay Harbour which later became the Bay Management Company. The BMC does not have financial autonomy. With such autonomy the Management Company might, for example, have used its funds for improving access to the harbour, or even providing housing. The Bay Management Company has a Bay Manager (appointed in February 1973) and pays one-third of the Harbour Master's salary (of which the Trustees pay two-thirds). With the development of the Secretary of State's commercial interests in the harbour, conflict with the Trustees became inevitable. Firstly, the Arunta/BOC base was established on Trustees' land, but a jetty had been built, without the Secretary of State's permission, into the Bay standing on the Secretary of State's seabed. Secondly, the Secretary of State had assigned himself to collect dues within the harbour without extinguishing the prior rights of the Trustees. In March 1973 a special meeting of the Trustees was called to discuss the problem of Arunta's jetty. The lease with Arunta clearly stated that all monies generated within the breakwaters would be paid to the Trustees, but the Secretary of State had intimated his intention of collecting dues under his 1972 powers. The question of double dues therefore arose.

The conflict of interest was resolved by the Secretary of State remitting the cargo dues (as for the ASCo base) and for the base to continue to pay the equivalent of the dues as rent to the Trustees. This settlement was arrived at after a considerable debate in the meeting of the Trustees and it was agreed to seek a letter from the Secretary of State preserving their rights under their lease with Arunta.

The next conflict arose over the use of the north breakwater; this entailed an exchange of vituperative letters between the Trustees' Collector of Shoredues and the Bay Manager. Again the issue was resolved in the Secretary of State's favour but with the cargo dues remitted in lieu of rent. At the time this was settled over £1,000 of dues from Arunta were written off in respect of the breakwater.

The only issue that seemed to unite the Trustees and the BMC at this time was their desire to recover overdue payments from Arunta.

The conflict between the Trustees and the Secretary of State is of some importance in its own right because it suggests considerable carelessness on the part of the Scottish Office both in failing to realise the full implications of the Secretary of State's assumption of power in the bay and in failing to explain and discuss the new situation with the Trustees. But the conflict is of considerably more significance from our point of view in its effects on the Trustees. To some of them it seemed as if the state was encroaching upon their historic rights and attempting to deprive them of income. Others, like the Chairman of the Trustees in 1977, took a pragmatic view; the Secretary of State probably had the powers he claimed and the Trustees were gaining income from cargo dues, so why argue with the Scottish Office? Informed legal opinion is that the Trustees probably had a very good case for maintaining their jurisdiction under Para. 81 of the 1873 Act. Furthermore the Trustees had a watertight contract with Arunta which the Secretary of State could not easily over-ride. Nor could the contract be over-ridden by saying that the contract operated outside the Trustees' jurisdiction. It is also the case that the Trustees control the land on to which cargoes are landed. This, in theory, makes the value of the jetty itself nil, and would provide BOC with grounds for seeking to reduce the very substantial rent it pays the Scottish Office. The point has not been lost upon BOC's lawyers. The Scottish Office did not particularly seek confrontation, especially as the Trustees were a public body and the appropriate body for spending cargo dues — the purpose of which are to maintain and improve harbour facilities. But to those Peterheadians who followed the legalistic line the pragmatists were betraying their trust.

This internal conflict was given a further twist after three Trustees were appointed to the BMC and it emerged that they and the Harbour Master had signed the Official Secrets Act. This was doubly insignificant in that all citizens are under the Act whether they have 'signed' it or

not and because it is a routine matter for all appointees to such government committees to 'sign' the Act. Some of the Trustees felt that they
should have been consulted over their employee (the Harbour Master)
signing and the Clerk was instructed to convey the Trustees' displeasure
to the Scottish Office. Their main fear was that they 'wanted to hear of
happenings in the Bay and not be met by a wall of silence'. In the
condition of incipient conflict between some Trustees and the BMC the
question of the Official Secrets Act exacerbated suspicion — was it
possible that the Trustees on the BMC were not telling all they knew to
the other Trustees? More importantly were they backing the Secretary
of State against the Trustees' interests in the BMC? The Clerk of the
Trustees (who was also Town Clerk) was also accused of selling their
interests by backing the Scottish Office; his actions may have appeared
this way at times but he was — in fact — bowing to what he saw as the
inevitable legalities of a situation which supported the Secretary of
State's position.

Put at its simplest: the Trustees had been absolute masters in the
waters around Peterhead and fiercely defended the rights of the fishing
community. The Secretary of State was now encroaching upon their
traditional rights; some wished to fight off this encroachment, others
wished to harmonise relations between the two. Those who served both
bodies were subject to the special hostility of those who wished to fight.
The conflict was probably exacerbated by the lack of any direct representation of the fishermen on the BMC Board. The Board comprised
businessmen and local government representatives; three of these were
Trustees, but none fishermen.

The Trustees still maintained their position on the question of jurisdiction and their independence from the BMC — with whom they
formed a liaison committee in 1973. When the BMC published its bye-
laws for the harbour, the Trustees informed the SEPD that any attempt
to enforce them within the Trustees' jurisdiction would land the SEPD
in court. The Trustees also raised wages in the harbour without consulting the BMC; this also raised the wages of the harbour controllers
who are used to supervise bay operations. The Trustees said that what
they did with wages was their own business; if the BMC wanted controllers they would have to pay the bill. The adoption of this stance
plainly created problems for Trustees who were on the BMC.

One particular example will illustrate the kinds of conflicts that
developed between Trustees. Because the Secretary of State had asserted
his rights over the bay it became necessary to amend the lease with

Arunta. The thirteenth clause referred to the cargo dues to be paid at
the rates leviable by the Trustees. But the dues now leviable in the bay
were to be decided by the BMC, so an amendment to the lease was draf-
ted to make this alteration. What should have been a routine alteration
to a lease became a battle because the amendment conceded, in writing,
the Secretary of State's rights in the Bay Harbour. A bitter argument
ensued in which those Trustees who were on the BMC, the Clerk and
the Chairman (who was Provost and strongly pro-development), supported the amendment against those who wished to resist the Secretary
of State, the BMC, the oil industry, developers and all whom were
thought to threaten fishing interests. According to those who attended
the meeting that finally approved the amendment the meeting was a
'shouting match' in which the dissidents were told they had to sign and
they replied that they would not be forced into doing what they did
not wish.

The question of the rent of the BOC base was still an open one. The
company was prepared to offer a fixed rent of about £20,000 a year,
but the Trustees preferred the old scheme of taking cargo dues, hoping
to make more money from prosperous years set against the poorer,
than from a fixed rate. Even the decision to discuss rents with BOC in
October 1975 was a source of argument within the Trustees' meeting.
The Chairman proposed to hold discussions against a motion to the
contrary, thus — in some eyes — aligning himself with oil and development.

In Chapter 3 we noted the Reporter's comment about the need for a
unified harbour authority. This suggestion is plainly highly contentious
in the current political climate around the harbour. The Harbour Trustees are not in principle against a unified harbour authority, provided it
is they (Minutes of the Peterhead Harbour Trustees, 8 April 1972). The
main fear of the Trustees was that they might lose control of the fishing
harbours through amalgamation and the unification of authorities.
Underlying this was the belief that both the Secretary of State and the
Regional Council (as the authority responsible for harbours) wanted to
acquire the very considerable income of the fishing harbours. In the
latter case the worst possible outcome would have been for Grampian
Regional Council to use this money to provide funds for smaller harbours within the region. Neither case seems substantial; the Scottish
Office had shown neither interest in nor flair for commercial activities;
Grampian Region could not have run the harbour on a day-to-day basis
but would have set up an independent harbour authority with powers

to spend its own income. The BMC, too, felt happy with existing arrangements and advised the Secretary of State that there was no reason why there should be one single harbour authority when they already worked harmoniously with the Trustees.

The crux of the issue was that there was already unified control of the harbour; although there were two harbour authorities, all movements were under the control of the joint Harbour Master. Detailed operations were controlled from one tower and there could be no ambiguities in the practical running of the harbours. The Trustees were also the sole pilotage authority and, although the organisation of pilotage was later adjusted, this fact too was unambiguous. The Reporter in making his comment at the Scanitro enquiry was echoing Shell who 'expected that a new statutory port authority would be set up in the future which would control all operations both in the fishing harbours and in the bay harbour'. In so far as both were concerned with safe harbour movements both misunderstood the arrangements already in force.

A second reason for opposing harbour unification exercised not only the Trustees but the Peterhead Business Association and others. This was the fear that at some stage the Bay Harbour would be organised under the dock labour scheme. If the harbours were unified, union labour would come to the fishing port. The Peterhead Business Association met with the Trustees and the BMC and the local MP was present at the first of these meetings. At this meeting it was said that dock labour could lead to the interruption of the smooth working of the harbour and the PBA 'recognise the importance of keeping the fishing industry in Peterhead viable'. At their next meeting, with the BMC, they suggested that dock labour would increase the cost per boat by £150 a week and thus take £½ million from the fishing industry annually. The BMC and Scottish Office solicitor disagreed: the dock labour provision excluded fishing ports and/or ports handling boats of under 80 ft in length. Peterhead fishing harbour would thus be exempt.

This contrary view, namely that the harbour would be exempt, is the view adopted by the SEPD and it does seem to be correct. The fishermen, however, were quite impervious to this interpretation of the dock labour scheme. Their reason for leaving Aberdeen was partly because of unions. The prosperity of Peterhead and the various harbour developments were felt to be based on their ability to unload their own boats. With unionised labour they would not only have to meet extra costs but possibly limitations to the times at which they could unload.

They also feared the loss of labour through sympathetic industrial action in support of dockers somewhere else on the northeast cost. Nothing would persuade them otherwise and if pressed they would say that the present dock labour scheme was the thin end of a wedge, and in the end all exemptions would be dropped. The Trustees in fact put the status of their harbours much more seriously at risk by loading grain ships in the South Harbour every autumn — they believed they were protected by the casual nature of this trade and its insignificance in the overall operation of the harbours.

In the face of such strong opposition the Scottish Office did not press the case for a single authority and on 13 September 1976 during a visit to Peterhead the Minister of State at the Scottish Office announced that there would be no single authority.

One other matter brought the Scottish Office and the Trustees into conflict. The 1873 Act established the Trustees as the pilotage authority. Peterhead is a compulsory pilotage harbour and dues are meant to be charged at a level sufficient to maintain a pilotage service. Fishing vessels are exempt from the pilotage requirement. In the relatively quiet and poor days of the harbours pilots had always been available but used rather randomly. Pilotage was *ad hoc* rather than compulsory. It was suggested by one harbour user that compulsory pilotage was enforced only to the extent that the dues were needed to maintain the pilot boat. This would not be untypical of a small harbour used mainly by small and local boats and a few regular visitors.

The coming of oil was interpreted by the Chairman of the Pilotage Committee as an opportunity to make money for the Trustees. This is quite contrary to the law; pilotage is a *service* and its accounts should break even every year. Also users of a harbour are entitled to representation on a pilotage committee, whilst the Trustees regarded it as their private concern. Officials from the Scottish Office visited Peterhead to 'sort out' the Trustees; the result was an enlarged committee with representation from BOC, ASCo and the BMC. The Authority had to provide a boat and 12 crew members to give 24-hour coverage, but commercial vessels needing pilots account for only 10 per cent of annual traffic. By 1980 the Pilotage Authority was in debt.

This issue illustrates how the coming of oil and the expanded use of the harbour brought about changes in the local community. By use and custom the Trustees had established rights which they were reluctant to see diminished. They, by usage, modified the law to suit their own conditions and customs. The new situation required dilution of their sole control, a more rational and orderly conduct of business in accordance with formal rules and law. After the intervention of a government department the local and informal system was replaced by one in keeping with the large-scale commercial use of the harbours by major operators.

Conflicts over the use of the harbour seem to have been fought with great vigour. Is it just a case of a bunch of provincial, narrow-minded and rather stupid fishermen fighting against change because they are agin any change? The scale of the conflict is explicable more in terms of interest than sentiment. There is much sentimental rejection of oil as something threatening a way of life — especially for those who see Peterhead as essentially a small fishing community. But Peterhead is a primary producer; essential to its economic activities are fish, fishermen, boats and a harbour. Given these then food-processing and transport provide employment for more than 1,000 people. But fish, fishermen and boats can go to almost any harbour. The special advantages of Peterhead's harbour bring the boats to its markets; the harbour is Peterhead's main means of production and employment. The livelihood of the fishermen and many others depends on the harbours and some feared that outsiders were wresting it from their control. It is an asset that can be shared but which Peterheadians would prefer to see in their local control.

For those who see little harm coming from shared control, and who believe Peterhead's future does not lie solely with fishing, the issue is not so acute. But all admit the legitimate interests of the fishing community. The points at issue have been the extent to which they can relinquish control without their interests being damaged.

The Trustees have done well out of the development in the harbour, the Scottish Office has not done well and may soon be seeking ways to dissolve its interest in the harbour — no doubt on the grounds that its sole interest was the national one of ensuring adequate supply bases for the North Sea.

Events in the harbour epitomise a problem we encountered throughout the study of Peterhead, namely that changes in Peterhead derive not from oil alone but from the changing fortunes of fishing. Some of the present Trustees can remember the Collector of Shoredues visiting a vessel on the slipway and asking for slippage dues in cash in order to pay harbour staff their weekly wages. Today they are doing well; in the year ending 7 November 1977 the income of the Harbour Trustees was £559,137 of which £59,993 — nearly 11 per cent — was from the BOC base. The Trustees are paying off their debts to the Scottish Office and to the old Town Council. (In this latter case they are withholding the last £63,000 until the Region and District have decided who should receive it as successor to the Town Council!) The Trustees want the District Council to have the money, as they fear that the Regional Council may put it into other harbours. During the fishing boom a new covered market had been built, the North Harbour entrance closed and the internal walls rearranged. Work was to commence on the covering of the slipway and the extension of the fish market, 50 per cent to be paid for by the Trustees themselves and the remainder jointly by the Scottish Office and the European Community. When this work had been completed the fisheries harbours would be developed to their limits, and no further expansion would be possible. The Trustees believed that in these circumstances, if there was a marked decline in the fishing industry, they could reduce their rates to the minimum required for care and maintenance and sustain or expand their share of fishing business. This would mean they would undercut all other ports, including some already receiving government support. When the loan was made for the new market the Trustees were asked by the Scottish Office if they would increase their very low charges, but they declined to do so. Peterhead fisheries harbours have therefore expanded to ensure their own future.

The Bay Harbour presents quite a different aspect. The Scottish Office — jointly with the NSHEB — made a substantial investment of £3 million in a jetty that was to service the power station, the ammonia plant and the NGL plant. The latter two will not now come and the NSHEB has reduced its usage from 75 to 37 estimated shipments per year. This reduction in traffic would lose the harbour £150,000 per year (at 1977 prices). If a substantial volume of gas or ethane were burnt in the power station the jetty would be used even more infrequently.

The ASCo base entails little profit: 100 per cent of the shipping dues and a percentage of the cargo dues are remitted to the company. The rent is the sole source of income and, although this rises over a 20-year period, the levels originally agreed did not allow for the rate of inflation that has been experienced and no review mechanism was built in. Even the rent from the BOC jetty and the breakwater could be substantially reduced if BOC were to contest the jurisdiction question. The future is uncertain for the Bay Harbour: a major offshore expansion would increase the traffic in barges and service boats; this depends on the profitability of the smaller North Sea fields. Offshore servicing could become concentrated in Aberdeen or it could be based in Peterhead. This would be even more likely if servicing activities attracted development and especially now that Aberdeen has lost full development status.

In the future ASCo may specialise in drilling supplies. BOC has plans to expand in the development and maintenance fields and with this in view has formed a joint venture with a firm of offshore engineers. But the major expansion of the Bay Harbour would have come with the development of petro-chemicals, which will not now come. The low level of activity in the still green fields around Peterhead is matched by the relatively low level of activity in the bay. (By 1980 bay traffic was 60 per cent down on the previous year and the £2 million BOC development was in cold storage.) Both depended upon developments that did not take place, both entailed heavy investment with limited returns. The Secretary of State for Scotland may well be glad to hand over the bay to the Trustees, providing the interests of other users are taken into account. The Trustees are unlikely to accept unless all debts in the harbour have been paid.

5 The community of Peterhead

A casual traveller arriving in Peterhead, looking around the town and maybe picking up a copy of the *Buchan Observer* might be excused for thinking that he had arrived in a true 'community' of seafaring folk. But the appearance is deceptive.

Peterhead has a distinct and isolated physical existence at the end of a difficult road at the eastern-most extremity of mainland Scotland. It gives the appearance of being a fishing community, an idea that is locally cultivated by Peterheadians. The fact that the inhabitants like to think of their town as a fishertoon was illustrated by Bealey and Sewel* by reference to a Sunday service televised from the town. The church was decorated with nets, floats and baskets for the occasion. The church, however, was one to which many of the town's elite belonged and had no ordinary fisher families amongst its members. The inhabitants underline their feeling of belonging to this community by reference to a common history and to a recent past when the fortunes of everyone depended upon the fishing industry. Peterheadians also refer to the qualities of independence and individualism derived from fishing and characteristic of the whole population. These qualities are exemplified by the resistance of many to accept social welfare benefits and opposition to trade unions.

The majority of Peterheadians speak a common dialect, a few surnames appear again and again in the town and like any small isolated population it is heavily intermarried. Every week the *Buchan Observer* reports on a multiplicity of voluntary associations and religious activity. The front page carries a weekly average of six advertisements for

*The Politics of Independence, p. 149.

activities based on churches or religious groups ranging from services to
the Full Gospel Businessmen's Fellowship monthly dinner and the Old
Parish Church jumble sale. The back page carries more such notices. In
addition there are about a dozen notices concerning the activities of
charitable, sporting, recreational and political groups. Sewel estimated
there were 60 voluntary associations in Peterhead in 1970, excluding
political and trade union organisations. The *Buchan Observer* promotes
the idea that Peterhead is a simple community of honest folk living
under the threat of outside forces, especially Mammon. It is, it should
be said, the only newspaper in the region that entertains the idea that
oil is not a wholly unmixed blessing – thus the folksy outlook and
archaic turns of phrase can be turned to sharp criticism. Is Peterhead
the classic 'little society' that has traditionally fascinated anthropol-
ogists; small, self-contained, relatively homogeneous with a shared oral
history and common culture? The answer must be 'no', or at least that
this is only a very partial picture of the town. Whilst Peterhead has its
own culture, it is not insulated from the wider economy and polity.

Peterhead began as a feudal property and became an object of speculation by 'robber merchants'. Control over many developments lay
ultimately in Edinburgh with the Merchant Company. The powers to
build and manage harbours depended upon parliamentary legislation,
and the ability to sell fish depended upon world markets and today –
in addition – upon EEC policies. Changes within fishing have resulted
in the use of larger boats and a consequential penetration of local share-
fishing by the large fishing combines. The harbours became heavily
indebted and in 1973 the expansion of the fish market was facilitated
by state loans and grants totalling £¾ million.

When Peterhead had its own Town Council it controlled day-to-day
matters of housing, street cleaning and refuse-collection, but planning
and education were under the control of the County Council. The economic base of Peterhead has widened so that it includes branch plants of
Nestlés (Crosse & Blackwell), General Motors and Thorn Electric
(Clarksons Tools). None of these is subject to local control and the first
two are foreign companies.

The social homogeneity of Peterhead is also deceptive. We have
already alluded to differentiation expressed in religious groups. Sewel
was able to arrange the voluntary associations in a hierarchy of prestige
which closely correlated with the residential location of the members.
There are posh and down-at-heel areas within Peterhead and posh and
down-at-heel people living in them. Social differentiation and social

placement is something at which Peterheadians are expert, and which was institutionalised in council house allocation to the extent of providing 'superior houses for superior people'. One of the clearest distinctions, however, is between fishing people and the remainder. Attitudes towards the fisherfolk are ambiguous: they are at the same time the salt of the earth, narrow-minded and greedy. As everyone's origins are known, social mobility does not always bring fully changed status; the local secretary of the Conservative party explained the apparently very tactless behaviour of a headmaster by commenting, 'What do you expect, he's only fisherfolk?'. The 'community' is also fragmented in another way and this too is evidenced every week by the *Buchan Observer*. Peterheadians have migrated to many parts of the world and every week the paper reports reunions, or publishes articles by 'exiles' or letters from people overseas seeking relatives in the town or Buchan area.

We will return to the question of social divisions in Peterhead later, but suffice it to say that there is no single homogeneous Peterhead community.

Recent events have impinged upon Peterhead and the ability of the population to control these events is extremely limited. Very little of my work has been concerned with explaining events by reference to factors indigenous to Peterhead, but rather with explaining the response of people in Peterhead to external forces. In doing this we are doing nothing remarkable; not even the late medieval city states were insulated from the wider society, and no town in Britain is free of the effects of world economic forces. No more was Peterhead an autonomous 'community' before or after the coming of oil, in spite of the *Buchan Observer*'s assertion to the contrary, namely that Peterhead was 'A once basically self-contained almost self-sufficient community' (25 July 1978).

Like most other populations, the people of Peterhead experienced the external forces partially mediated by the activities of the state. The state, then, is one of the external forces whose actions impinge upon local populations. The state is only one agency which exercises social control within a society. It controls with the backing of law and is unique in being able legitimately to resort to the use of force. The state is not simply one institution but an ensemble of organisations concerned with government at national and local levels, with the provision of public services and utilities, with 'law and order' and the waging of war. The state is an object of very considerable interest to social scientists and analysis of its functions is very central to current debates within Marxism. Our main interest is not in the abstract notion of the state but in the effects of government upon the outcome of events in Peterhead. Chapter 6 is devoted to a consideration of the practical effects of recent oil-based developments upon Peterhead, but in order to make sense of these effects we need to understand the impact of policies developed outside Peterhead in modifying or constraining the actions of the individuals and agencies involved. Most obviously Peterhead lacks and has always lacked direct control over national economic policies controlling wages and prices; the central government and provincial administration devise regional development policies and the Regional Council oversees planning with the District Council working out detailed local plans. Although it might be consulted, Peterhead does not control policies such as these.

Prior to local government reorganisation Peterhead had a Burgh Council of twelve councillors from whom were selected magistrates and representatives to the County Council. In common with the practice of the small Scottish burghs the County Council representatives were expected to 'get on with the job' and not to report back to the Burgh Council. The councillors were elected on the basis of three-yearly re-election, with vacancies filled by co-option between elections. The councillors did not represent political parties and the most striking feature of Peterhead political life was the absence of parties. This had direct consequences for the conduct of council business because the councillors adhered to the idea of Peterhead as a unitary community in which the people's representatives could agree by a process of rational discussion — even when there was no rational basis for agreement.

Bealey and Sewel's study of the politics of Peterhead conducted from 1968 to 1971 dwelt upon the conduct of council business in some detail. I need only summarise this and add that I discovered nothing that did not support the Bealey and Sewel findings. The lack of parties meant that there was little organisation of council business and the outcome of a debate could not be foreseen because a majority would not be whipped in to vote. In these circumstances, firm control of the conduct of council meetings was essential so that clear decisions could be made in the meetings. According to Bealey and Sewel this was not the case; there was a marked lack of control. The discussions rambled on with no clear motion and amendment, unseconded amendments would also be discussed and councillors were often at a loss to understand the details of the decisions they were making. Usually the Town

Clerk and the Treasurer were drawn deeply into the debates so that they seemed to be acting with the status of elected representatives. Given the lack of clarity, it is entirely understandable that these officers should be drawn into unravelling the discussions — the outcomes of which they had to implement.

To abstract the sense of a meeting was naturally very difficult, so the minutes did not always contain the decisions that councillors thought they had made. Therefore acrimonious arguments developed over the minutes in which councillors even appealed to the press and the observing sociologists to sustain their case. What was happening was that after protracted debate each thought they were all agreed upon what they had said. It was impossible however, to formulate a simple resolution embodying the alleged agreement because that would have started debate again.

In these circumstances it was extremely difficult to draft minutes which provided any basis for action: whichever course was taken would have pleased only a minority.

The *Buchan Observer* and other commentators (including the ex-councillors to whom we spoke) seemed to agree that the quality of representatives was poor and had deteriorated. No ex-councillor was unwilling to abuse other ex-councillors in private to the investigator, and malice seemed to have abounded. The lack of 'quality' could be observed from time to time when ex-councillors made elementary errors in discussing local politics: for example asserting that the burgh was its own planning authority (a common belief amongs ex-councillors) or that industrial developments would increase rateable values and keep rates down (in fact it results in a *pro rata* reduction of Rate Support Grant). Yet the councillors adhered to the view that Peterhead was a homogeneous community and that fundamentally they agreed with one another. Their 'agreements' were confusing and sometimes spurious because they were really only agreeing that they ought to agree, not agreeing to a substantive issue. Paradoxically contention and criticism was not welcomed in council, and critics, according to Bealey and Sewel, were dealt with by being made convener of the committee whose work they were criticising. Thus a councillor who was known to be a Labour supporter was made convener of the housing committee — in the added and misplaced hope that this would weaken his electoral position by making him the focus of public criticism.

The Bealey and Sewel study also showed that the population of Peterhead knew their councillors but that they found the work of the council 'shrouded in obscurity' — a feeling which may have been shared by some councillors. Some of the respondents to Bealey and Sewel's survey said that they followed the affairs of the Westminster government more easily because there was more publicity about its doings. Those who lacked political understanding in the case of national politics blamed the mass media but if they failed to understand local politics it was because the council were 'under-handed' and secretive.

Nearly a half of their respondents had little comprehension of the working of the Peterhead political system even to the extent of knowing how to get elected. As we will see below, the Burgh Council had a high rents and low rates policy, less than 1 in 5 knew of the former and 1 in 4 of the latter, only three respondents knew both to be the case and about one-half thought that Peterhead's rates and rents were 'about the same' as the rest of Scotland. Later in their questionnaire respondents were asked what they would have thought if (hypothetically) high rents and low rates had been the case: 44 per cent thought it would be wrong or unjust.*

From Bealey and Sewel therefore we gain a picture of a not very expert council and a relatively ignorant electorate. Two questions arise: (1) how was this possible and (2) did this affect the course of events in the development of the impact of oil?

The basis of politics

It was commonly argued that 'politics should be kept out of local government', but Peterhead would probably have been better served by political parties. The parties would have defined issues clearly, related them to national party policies and taken steps to inform the public about issues even if only so far as was necessary to win their votes. It is inconceivable, for example, that Peterhead's high council rents and low rates (amongst the highest and lowest in Scotland) would not have been a political issue had there been a Labour party in the town. Why then was there no Labour party in Peterhead? Quite simply because there was no base for such a party. One such base would have been industrial; but the main industry was and is not unionised (fishing) and nurtures strong free-enterprise anti-union sentiments. The main industrial base would have been the organised male workforce of the two engineering

*The Politics of Independence, p. 186.

works, but there the union officials were too overstretched on industrial matters, actually building up union strength, to be able to devote the time necessary to building a political movement. The Labour party in Peterhead, in fact, gained its greatest momentum when industrial and political issues coincided in the 1971 Industrial Relations Act.

The second possible political base would have been amongst council tenants who constituted a major interest within the town. A political activist could build his constituency upon housing issues, neighbourhood politics and community affairs in working-class districts. But councillors were elected at large, for the whole of Peterhead, and no councillor could run for election openly representing the interests of council tenants because they constituted only 45 per cent of his potential electorate. Such a 'council house candidate' could be represented as threatening the owner-occupiers with higher rates.

Thus working-*class* interests were not mobilised in the Peterhead political structure. But the class interests of property were, and this was embodied in the low rates/high rents policy. In talking to ex-councillors, I was very struck by the depth of hostility to council tenants, rooted in most cases in the belief that people should be independent and own their house, but in a few cases rooted in what one can only describe as outright hatred. Excessive measures against council house tenants were likely to bring the council into disrepute and it is in this respect that the Town Clerk may have used his skills in organising council business in never letting the suggestion that council tenants should be evicted for untidy gardens ever come to full debate and decision.

The minority of councillors sympathetic to Labour and the working-class interest reported many cases of other councillors attempting vindictive measures against council tenants who were sometimes in minor breach of council regulations (like a daughter 'overcrowding' her father's council house whilst caring for him in sickness). Housing was the most socially significant item over which the Burgh Council had direct control and Bealey and Sewel found 'it is impossible to escape the conclusion that there was a widespread suspicion about house letting, transcending all groups . . . .' This mainly derived from the informal zoning of the town into more or less rough and respectable council house areas and widespread allegations of favouritism and canvassing by councillors in making personal allocations of housing. As one ex-councillor put it, 'some councillors felt that this made them somebody to keep people waiting outside their home on a cold night' canvassing for a house. This Labour councillor, when housing convener, reformed the allocation system to give much more power to the Housing Factor because he reckoned the personalised allocation scheme was 'corrupt', a belief for which he offered some supporting evidence in terms of councillors having favourites and doing 'favours'.

We have suggested that the working-class interest has not been organised in a Labour party in Peterhead, and yet class politics seemed to be important, certainly as evidenced by the one sphere in which the council had influence over people's lives, namely housing. Was there an anti-working-class party? Again the answer is negative; the Conservative party, for example, is not particularly strongly organised in the East Aberdeenshire constituency, largely as a result of Bob Boothby's incumbency of the parliamentary seat. He discouraged party organisation and instead mobilised his support through personal contacts. It was observed wryly by a number of informants that the Conservative party was never so well organised as to get rid of Boothby's successor, Patrick Wolridge-Gordon.

A number of Burgh Councillors were members of the Conservative party, but they did not use this label openly in local politics. At reorganisation two Burgh Councillors, T. J. Smith (the last Provost) and Mrs Lamb, were forced to take an open party label in order to participate in the more ordered business of the Regional Council to which they had been elected. Both took the Conservative ticket and both were beaten by SNP candidates in 1978 local elections (in which no party offered any policies). In so far as it is possible to ascertain the party political allegiances or sympathies of the councillors immediately prior to reorganisation in 1975 the Burgh Council comprised six Conservatives, two SNP, two Labour and one Liberal.

What, then, was the political base of this old Burgh Council? The councillor's electoral base can be understood by reference to social networks in Peterhead. Voting was very much on a personal basis and individual councillors appealed to the 'British Legion vote' or the 'churches vote' according to the pattern of their personal activities and acquaintances within the town. These kind of allegiances would have been undermined by introducing 'politics' (i.e. parties) into local government. The councillors were also linked to one another and to other influential people by common membership of a small number of voluntary associations. It is the linkages between associations rather than the social position of any one member that is important politically. Of the 60 voluntary organisations identified by Bealey and Sewel, 6 were identified as being especially closely interlinked and of high standing in the town. These were: Rotary, Round Table, Professional and Business-men's Club (a lunch club) and three churches (two of only 'medium' status according to Bealey and Sewel). When I linked these associations with local employers and managers and the Town Council we found a very close network that might constitute a local elite. In Figure 2 the interconnectedness of associations is indicated by the breadth of the lines joining them and the positions of Burgh Councillors are numbered.

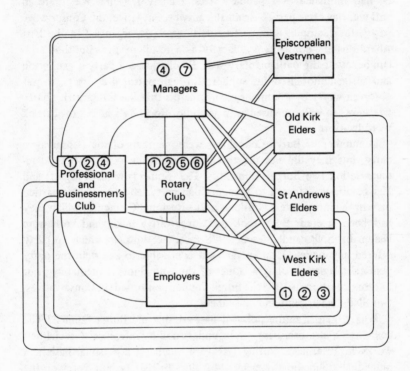

Figure 2

Thus, whilst managers, for example, were not all councillors they did meet councillors at Rotary and the Professional and Businessmen's Club or were elders of the same church. Conversely there were many situations in which councillors would meet members of the business community, if they were not already themselves businessmen.

If we add a second set of relationships connecting the Harbour Trus-tees and the Council to these associations we can see in Figure 3 just how densely the Council and Trustees were connected with proprietors and businessmen through this network of voluntary associations.

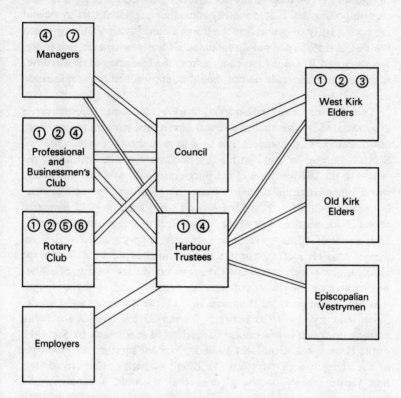

Figure 3

Party activists and trade unionists were connected only very loosely to these networks, one party activist and one trade union officer only served on the council. The members of this network were further but more sparsely connected through participation in a number of local charitable organisations. Crucially, the majority of members of these associations were non-manual workers and, of course, the first three listed were exclusively associations for businessmen.

The town's leadership was therefore drawn from business and property rather than labour. In the late 1960s, when there was some public discussion of the poor quality of the elected representatives, the PBC,

Rotary and Round Table discussed the question of recruiting better candidates. The outcome seems to have been that three members of the Round Table ran successfully for election, helped by other members' cars on polling day. Later another councillor suggested that in view of the poor quality of councillors the Burgh Council itself should approach the PBC and Rotary to put up candidates. Clearly, whenever 'leadership' was discussed it was seen as coming from the businessmen; it was never suggested that the trade unions should be approached to provide candidates.

The dominance of non-manual workers and small proprietors is demonstrated by the council's almost obsessional concern to keep rates down. Peterhead demanded rates that were in the 16 lowest in the 175 Scottish burghs, in other words the rates were in the bottom 10 per cent. For the last five years of the Burgh rates were at 63p in the pound whilst other burghs of comparable size charged from 84p to 107p. Rateborne expenditure in Peterhead was £18 per head in 1970 compared with £36 for Scotland as a whole.* The effect of this was that whilst the basic needs of lighting, street-cleaning, water supply and sewage disposal were adequately met, the town lacked any further provision. Youth, for example, was poorly served and the provision of parks or other facilities was resisted on the grounds of costs and further justified on the grounds that nature had provided Peterhead with excellent beaches and the lido. More seriously, by 1970 between 20 and 25 per cent of the entire housing stock was below the standards defined as tolerable by Section 2 of the Housing (Scotland) Act 1969. Perhaps the most extreme example of the concern with rates was the Burgh Council's failure to adopt a rent rebate scheme on the grounds that it would undermine older people's sense of independence. This is also an interesting example of the way in which aspects of local culture can be mobilised for ideological purposes. All the ex-councillors interviewed stressed the importance of keeping down rates, although the Labour ex-councillors were on the whole very critical of the policy. Latterly the loan repayment from the Harbour Trustees was used to pay rates rather than provide additional amenities. Interestingly this was described to me as an example of how the Harbour and the Burgh stood and fell together but this was just an example of the ideology of community solidarity obscuring the reality

---

*I am indebted to John Sewel for this information. Keeping the rates down is a policy objective of great antiquity, according to Bealey and Sewel. The whole of the first part of this chapter depends very heavily upon their work.

of a bitter dispute between Council and Trustees over the repayment of the loan. The town was probably under-rated in the last years of the Council and handed a deficit to the new Regional Council, although the facts of this are hard to unravel and hotly contested. When rates trebled after reorganisation the ex-councillors attributed this to the cost of large Regional and District bureaucracies, and the Region attributed it partly to under-rating prior to reorganisation. Rates were, therefore, a central issue in Peterhead and keeping them down was more important than providing some of the amenities that would have enhanced the town. Even more importantly low rates were regarded as a higher pri- ority than either keeping council rents down or providing more council housing to relieve local need.

In many respects these considerations are largely of historical interest because Peterhead ceased to have an independent political existence in June 1975. But our study is in part historical and we have sketched an answer to how the politics of Peterhead 'worked' without parties. The next question is whether the peculiar political structure of Peterhead had an effect on the way in which oil impinged upon the town.

The political impact of oil

Sociologists of development have been especially interested in the role of those who mediate the influence of incoming corporations. These mediators, or brokers, are usually drawn from the native middle class or political elite. They are valuable to the incomers because they wield power locally and can get things done within the local community. They also have a fund of detailed local knowledge and local contacts. Whilst they are valuable to the incomers, they are not indispensable, certainly not after the first penetration has been effected, but they may be useful for local public relations and as evidence of the corporation's 'local' dimension. The broker himself stands to gain; possibly economi- cally through fees, retainers, or a salary. If he is a businessman he may also gain business for his own enterprise or closely associated enterprises. The broker may also gain politically by representing a powerful com- pany and his powers of local patronage are increased through his ability to direct business to local firms or to bring local people into the pres- tigious network of the incomers. The broker comes to represent the incomers to his native community and in becoming their 'man' he may become detached from the local community and be seen to represent incoming interests *against* those of the local community, at which point the broker's usefulness to the incomer is much reduced, unless – as is sometimes the case in the third world – the broker can be backed by force (physical violence) or wider national and international political power.

The men who stand between multi-national corporations and the native society and whose wealth, power and prestige is dependent upon these foreign rather than local interests, are collectively known as a 'comprador bourgeoisie'. Did a comprador bourgeoisie develop in Peterhead?

We can tackle this question by asking what such a class would have to offer incoming interests. The answer is, very little. The Burgh Council, for example, had very little control over the outcome of issues that concerned oil-related interests: ownership and zoning of land, planning permission, provision of grants, loans and services. These all lay within the ambit of the central government or the County (later Regional) Council. Some of the local councillors and businessmen could, however, offer local knowledge and others could lend their locally based prestige in supporting the incomers' projects at County and Regional level.

At the beginning of the oil-based boom there was only one active and effective law firm in Peterhead and inevitably the early arrivals in town went to this firm for expert advice and professional services. The firm advised Arunta and the speculators, Site Preparations and Peterhead and Fraserburgh Estates and became involved in a very wide range of land transactions entailing the acquisition of options and the actual sale and purchase of land. One partner was especially favourably disposed to economic development in Peterhead and also helped the incomers professionally in their dealings with the various local authorities and the Scottish Office. There is no doubt at all that if there was an award for the most unpopular man in Peterhead, this partner would have won it. The unpopularity was, however, tinged with admiration because he had taken the 'main chance' and done well out of oil. He was a 'wily businessman' and 'crafty', for which he was admired and to some extent envied by some of the business community. He was also the only person in Peterhead who directly refused to talk to me.

A second obvious candidate was the Provost, T. J. Smith. He was not only Provost but a member of the County Council and the Harbour Trustees. He was also strongly in favour of development. He certainly countered the opposition to Scanitro and used his vote in favour of Arunta at meetings of the Harbour Trustees (see minutes of Harbour

Trustees, 6 May 1972) and of the Town Council (Minutes of Burgh Council, 13 November 1973 and 14 January 1975). But according to one incomer, as Smith was absolutely honest and totally without imagination, he could then be neither corrupted nor inspired. He was therefore useless to this informant. As a proponent of development Smith probably smoothed the path for the developments that took place, but he did this because he believed it to be in the town's interest. The significance of this is easily seen when one considers the likely effects of an anti-developer being Chairman of both the Town Council and Harbour Trustees. His ability to defer and delay developments could have driven developers away.

One person who was widely accused of 'going over' to the oil interest was Ian Craig. He was a burgh councillor, Treasurer of the Harbour Trustees and a member of the County Planning Committee. In theory he had a great deal to offer an incoming developer; good local elite connections both in the business and fishing industry, detailed knowledge of landholding around the fishing harbours, fore-knowledge of the county's planning policy and the ability to influence it. He was employed by Clarksons Tools in administration but moved to personnel management in Arunta, which he later left. Craig may have helped Arunta establish itself in Peterhead, but there is no evidence that any of the information to which he had privileged access would have been of any use to Arunta, or that he gave it to them. A list of landholdings around the harbours was circulating in Peterhead but it did not originate with Craig. The main outcome of Craig's contact with oil interests — specifically Arunta — was an advancement of his own career by taking new job opportunities when they offered themselves.

The remaining support from burgh councillors was indirect and not always unconditional. One felt that the exploitation of oil was in the national interest and that Peterhead should not block this. Mrs Lumsden founded the anti-Scanitro campaign, but changed her mind after the Pogsrunn visit, but still opposed environmental degradation and waste. Councillor Baird, a Labour councillor, supported oil-related developments inasmuch as they created jobs and providing the necessary social and service infrastructure was provided and not paid for locally. Councillor Alexander opposed oil developments on safety grounds but was 'all for' extra jobs. Similarly Mrs Lamb was in favour of more work but worried about the social effects of affluence on a community in which people no longer sent their children to Sunday School.

The most significant fact about the proponents and opponents of oil related developments is that the former had little to offer incomers, beyond professional services, and the latter could do little to prevent developments.

One other grouping within the town was not mentioned by Bealey and Sewel because the Peterhead Business Association was not founded until 1973. It described itself as 'a non profit-making co-operative organisation formed to respond to the challenge of the new developments in and around Peterhead.' Some of its leading members seemed impatient with the relatively unenterprising response of local businesses to the developments taking place. Its objectives were:

(1) To safeguard and represent the interests of Peterhead businesses.
(2) To provide the maximum opportunity for existing businessmen to expand their businesses.
(3) To provide a medium for the exchange of commercial information.
(4) To encourage co-operation on a sound commercial basis between existing businesses and new businesses in Peterhead.
(5) To safeguard the long-term commercial future of Peterhead.
(6) To do anything calculated to improve the commercial future of the town.

Membership was by business organisation, not individual. In discussion with leading members of the Association, it was clear that some of them saw the role of the PBA in Peterhead as similar to that of the North East Scotland Development Agency (NESDA) in Aberdeen. In fact NESDA was a Regional Council agency intended to serve the whole region but in Peterhead it was regarded as an agency that directed developments to Aberdeen and 'no further north than Dyce'. The PBA would, according to this definition, be a development agency encouraging firms into Peterhead in order to 'improve the commercial future of the town'. As such it would comprise an intermediary organisation connecting the local community to incoming commercial interests and facilitating the penetration of the local economy by the incomers.

The PBA did not, in fact, operate in this way and has had very limited success in encouraging new business to Peterhead. It has been concerned not just with oil developments but, in accordance with objective (5), the encouragement of long-term developments that would outlast oil. Objectives (1) and (3) proved to be the most readily achieved and the PBA became a successful pressure group representing local business. Between 1973 and 1975 it took an active interest in questions of demand for local labour and the need for housing and ensured that these two items were on the agenda for the local authorities. In collaboration with the Harbour Trustees and with the local MP attending one of their meetings, they considered the Shell proposals and stressed the need to keep the fishing harbours open at all times. In 1976 they took up the question of reopening the railway line to Peterhead and the possibility of an amenity fund provided by the oil industry to make sure that the locality benefited directly from oil and did not lose through reduced Rate Support Grant.

An issue upon which the PBA was again at one with the Harbour Trustees was that of opposing the unification of the harbour authorities. They found themselves in this position because it was believed that unification would lead to unionised dock labour and therefore unacceptably increased costs to fishing. Here we see the PBA taking a strong stand against the interests of the Regional Council, the hopes of the Scottish Office and the preference of the oil industry. As we have seen in Chapter 4, that pressure resulted in an agreement not to unify the harbours. The pressure group activities of the PBA in concert with others seem to have been sucessful.

They were less successful in trying to attract oil business to local firms by persuading incoming businesses to use local suppliers. The reply they received was that this would be done if local firms were competitive; in other words no special consideration would be given to local firms.

One other interpretation of the PBA's role was that it could fill a gap left by the demise of the Burgh Council. This role became quite clearly defined from about the beginning of 1977 when the PBA's minutes show a shift from discussion of wholly commercial issues to questions of town planning, traffic and environment. The PBA not only made strong representations on these questions to the BBDC Director of Planning and the highway authorities, they were eventually consulted by both Grampian Regional Council and BBDC, which was as much as the old Burgh Council could have hoped for. None the less this represents a marked shift of emphasis from a 'development agency' to an 'amenity group'.

The PBA remained active but its effectiveness in achieving its aims probably depended mainly upon the informal contacts it sustained between businessmen and only partially upon its role as a pressure group. It is difficult to judge the effects of such an association but it is probably fair to say that it has enabled and facilitated the most progressive elements amongst the businessmen in Peterhead to adapt to the changing circumstances and perhaps to profit from them. What it does not seem to have become is an organised bridgehead for outside businesses or new developments in Peterhead.

With the demise of the Burgh Council in 1975 the town became simply a part of the Banff and Buchan District. According to the *Buchan Observer* and ex-councillors the effect of this was entirely negative. Peterhead's new road-sweeping machines and refuse carts were 'taken away' by the District Council and the present littered streets are the result (in the 'old days' when Mrs Lumsden complained about the untidiness of the streets she was made convener of the Cleansing Committee). It is widely felt that Peterhead's interests are no longer properly taken account of. This seems to be largely a response to the loss of *local* control rather than a comment upon the quality of government. One of the most radical changes following upon local government reform has been the institution of participation exercises in physical planning. Thus the Roanheads development, a town-centre revitalisation scheme, was conducted with consultation with the residents and a shop was set up as a centre for enquiry and comment. Peterheadians did not respond very enthusiastically to this, but those who did were quite amazed to find that their opinions were sought and listened to. This is in sharp contrast to the Burgh Council who *as a matter of policy* did not consult the public, on the grounds that they were elected to get on with the job of governing.

The loss of the small measure of political autonomy enjoyed by Peterhead was not the result of the suppression or domination of local political interest by multi-national corporations but the result of reform unconnected with oil developments. Given the promise of quite drastic changes resulting from oil, however, local people did look for substitutes for the old Burgh Council. According the the *Buchan Observer* the proper body to represent the town was the Feuars Managers. According to some Harbour Trustees the most representative body was the Harbour Trustees.

Peterhead was more acted upon than active politically from 1970 to 1978. People had few resources with which to bargain with incoming interests. The harbours were one, and these they defended very effectively, but it was not the Burgh Council that did this but the users of the harbour and their allies. A second resource was land, and this was not collectively owned by the town and was entirely vulnerable to speculative activity, as we have seen. Control of the *use* of land lay at county, regional and district level. Thus Peterhead had no important and independent political role to play and, furthermore, lacked the resources that could have been 'traded' to provide the power and prestige basis of a new local elite dependent upon the oil industry.

In sketching the lack of autonomy of Peterhead, I have implied relatively stronger sources of political power outside the town. In common with every other town in Britain, Peterhead is subject to the authority of *the state* in its various manifestations. The policies adopted by the state are usually oriented to national considerations and how these impinge upon any one locality is a matter for empirical enquiry. Three aspects of state policy have been especially influential upon Peterhead, those policies concerned with the economy at large, with regional development and with planning.

Economic policy

An incomes policy of some kind has been in force from the beginning of developments in Peterhead. The wage increases that employers might have given have been limited and to some degree collective bargaining has been restricted thereby. In Peterhead this resulted in what local employers saw as unfair competition because an incoming employer could leap-frog over the existing local wage rates in order to recruit the labour he needed while the local employer was unable to respond by raising his wages. We know very much less than we would like to know about wages in Peterhead. Some general features may be noted. Average wages for male manual workers in Scotland rose from about 96 per cent of the UK average to 102 per cent between 1970 and 1974* and from somewhat lower previously. But there are more important issues, has the gap between the highest- and lowest-paid worker widened significantly, has a two-tier wage structure developed, is there a dual labour market? These are no less important questions in Peterhead where some have left local jobs to join the oil industry. We know very few of the answers: in 1977 when the national average wage was £78.60 just over 13 per cent of Grampian Region's male workers earned less than £50 a week.† The gap between them and men working in the more highly paid oil jobs must have been widening and the public discussion of oil bringing 'big money' to the northeast must have seemed especially ironic and frustrating.

*M. Gaskin *et al.*, *The Economic Impact of North Sea Oil on Scotland*, HMSO, 1978.

†F. Twine, 'Scotland', in *Low Pay in the Three Nations.* Low Pay Unit Bulletin No. 25, 1979.

Just how disabling was pay restraint for local employers? This will be discussed more fully in Chapter 6, but we found evidence that pay restraint was a contributory factor in preventing some local firms expanding. But loss of labour may not be attributable to wage restraint alone: for example, in 1973 a lot of local firms and the prison lost an abnormally high number of employees, but as we will see these were largely skilled tradesmen who had not been working at their trades for lack of opportunity. Construction work provided many with a chance to practise their trades and they might have done this without reference to pay policy. In addition the construction industry and the supply bases provided the possibility of overtime and bonuses. It is unlikely that local employers could have competed with this in any circumstances. I also concluded that pay policy was easily evaded and that local employers could raise their employees' incomes (although not necessarily their wages) when they wished to do so, and that some of them did therefore compete with oil-related industries. There was, however, local resentment in December 1973 when an investigating team arrived to make enquiries into alleged infringements of the pay code. Those who did not meet 'unfair competition' in this way and others who lost workers found ways of adapting to new lower levels of manning as we will see.

A second area of economic policy that impinged upon Peterhead was cuts in public expenditure. The purpose of this measure was two-fold and wider than any local Peterhead considerations. Firstly, the cuts reduced spending by the government in order to redirect more money into private investment, but secondly the cuts reduced standards of living by reducing personal consumption of services — whether these were education, the use of roads or family planning services. These cuts had broad consequences for the region but some were especially problematic. For example, one response of the Regional Council was to make no additional appointments to its establishment. Yet the problems created by oil and oil-related activities made extra staff very necessary. This is very clear in the case of planning and research: a great deal of information was required about labour migration, journey to work and wage levels in addition to answers to a wide range of technical questions about pipelines, oil installations and chemical hazards. The research needed to provide a data base for social planning was not being carried out as effectively as it might have been.

The most serious loss in this respect was the cancellation of the 1976 Census as part of the public expenditure cuts. The population of

Peterhead in 1971 and 1981 is known but the simple, total number of people in Peterhead in 1976 is not known. Yet it was in this mid-period of the decade that temporary labour was at its maximum and when possible long-term changes in the population might have been taking shape.

The lack of elementary data can create great difficulty for educational planners who have to erect buildings without knowing if there will be children to occupy them – a problem driven home by the *Buchan Observer* in a sarcastic report of a meeting at the New Academy, published on 4 October 1977 under a typical heading 'Boom-boom Peterhead's Cuckooland Accommodation'.

The work of various 'social work' agencies, departments of social work, DHSS, ESA, guidance teachers and voluntary organisations generated data that might prove useful in assessing the social problems that the Reverend Miller and Dr Taylor thought would result from developments in the town. These data have not been brought together in a useful way because there are no resources with which to do so. The records of one local government department were in such disorder that they had generated two different sets of data with which different officials within the agency were working. It was a small department which had had to cope with reorganisation of the professions involved, local government reorganisation and oil-induced changes. Part of its data problem had been solved by the destruction of records. A case could easily have been made for the employment of two or three social statisticians and sociologists, who by reviewing and collating existing data could contribute significantly to planning and policy formation in the district and region.

Inability to solve the problem arose as much from the administrative details as from the nature of policy itself – but the mode of implementation could have defeated the goals of policy. Funds were available from central government for extra work undertaken in connection with oil developments but the money was paid in arrears, so the only way this source of finance for extra research staff could be used was to reduce staff elsewhere. Lack of social research was a major concern to officials and one of a number of hindrances in policy formation.

Development policy

In the early stages of capitalist development the resources of the countryside were appropriated for the development of the industrial towns and cities. This was not solely a polarisation of industrial and agrarian society. In the English provinces, for example, Durham County had its main resources exploited in the nineteenth and early twentieth centuries, and the profits from this were invested elsewhere in banking, property or new industries or invested overseas. Thus a county which generated considerable industrial wealth was by virtue of its social relations to capital deprived of benefit from that wealth. Industry and the state became centralised and the policies of the modern welfare state were initially concerned with the amelioration of conditions in the towns and cities caused by rapid economic expansion and population growth. Social policy developed therefore out of a concern with urban, metropolitan issues.

At a later stage of development the peripheral areas may again become useful as sources of new raw materials or labour. Voluntary relocation is one strategy open to a company seeking to reduce labour costs, another is capital investment in machinery that will replace men, others are importing migrant labour or employing women. Thus after the First World War (when they relocated for strategic reasons) Crosse & Blackwell found themselves with a cannery in Peterhead in an area producing fish and agricultural products that they could export in tins, and a large pool of female labour to be drawn from. General Motors and Cleveland Twist Drills were persuaded to come to Peterhead because suitable labour was available. Recently in a rather different context, because geographical choice was more limited, Peterhead North Sea Terminus was able to advertise non-union labour to potential developers.

The state too has policies for the peripheral areas. Most obviously these are regional policies designed to help identifiable areas, but there are also policies more generally concerned with counteracting the uneven distribution of wealth, income and social capital. Rating policy is a case in point; the Exchequer operates a scheme for balancing the rates between richer and poorer authorities. One of the major claims of those who sought to encourage development in Peterhead was that projects such as Scanitro, NGL plant and the bases generated high rateable values and therefore high rate incomes for the locality. This is true in that higher rate income would be generated but spurious because a *pro rata* loss of Rate Support Grant would keep the net income the same. In this case therefore arrangements to help poorer areas also prevent them from benefiting from economic development by redistributing the increased rate incomes away.

Regional policies as such comprise a series of attempts that have been made over a long period to reduce the imbalance between the under- and over-developed regions. The reasons behind the evolution of such policies do not concern us here, but only the outcome of the policies which affect Peterhead. Peterhead is part of the Grampian Region which until 1979 was part of the Scottish Development Area. Incoming companies, therefore, qualified for any of a range of grants, tax allowances or loans. The grants and loans were as follows:

Regional development grant. This covered 20 per cent of the building cost of new plant and machinery. It was available for enterprises in manufacturing and processing.

Removal grants. Projects moving into a development area might qualify for up to 80 per cent of the cost of removing plant and materials and the employer's net statutory redundancy payments at the previous location.

Selective investment schemes and *Interest relief grant.* The former was confined to projects of at least £½ million but both contributed to the cost of interest charges, at a level to be negotiated.

Transferred workers removal assistance. Under this scheme grants were made either for moving key workers in, or for sending local unemployed for training at a parent plant. The grants covered travel, removal, disturbance, etc.

Industry schemes. Certain industries (including machine tools, wool textiles, red meat slaughter) qualified for grants and loans for building and machinery needed to modernise or rationalise production.

Service industry grants. These covered a grant of £1,500 for employees moving to the area, a grant of £1,000 for each new job created and a grant to cover rent for 5 years. To qualify the firm must have created at least ten new jobs and it must have had a genuine choice of location.

Tax allowances. In the first year in a development area an enterprise might claim 100 per cent of capital expenditure on machinery and equipment and might write off say 54 per cent of the construction cost of buildings (and then 4 per cent per year).

Incoming industry therefore seems to attract very considerable state subsidies in a development area. It was to this that Michael Smith was alluding in the planning enquiries. Because the whole of Scotland was a development area, wherever Scanitro or Shell/Esso located themselves they would at least have received 20 per cent of their capital costs from the state, 54 per cent of these costs allowed against tax and the whole of their equipment costs in the first year. The most that Smith and other local employers could have hoped for was a cash grant or loan towards modernisation. It seemed unfair that local employers had weathered economic difficulties in the local community and were 'loyal' to Peterhead only to find that a newcomer after rich pickings could gain state aid on a big scale.

A sociologically significant point about this array of grants and loans is that there is an implicit theory underlying it. It is a very simplistic theory of economic development based upon the classic development of industrial capitalism. The development areas are either 'backward' and therefore need modernisation (through industry schemes) or in need of 'development' and this is equated with the growth of manufacturing or processing. Why else exclude service industries unless they have a genuine choice of location? Obviously oil-related service industries do not have this choice, they are where the oil industry needs them. So perhaps the logic is: if an industry will come anyhow, why subsidise it? The same logic could be applied to Scanitro and Shell; they would have been sited in a technically and commercially suitable location in order to carry on a highly profitable activity. Should they therefore have benefited from the inducements offered to encourage other companies who might otherwise not have chosen to come to a development area?

The logic of regional policy seems not to fit Peterhead easily and it was not policy devised to cope with the situation then found in the oil-affected areas. The servicing and maintenance of offshore installations will be a long-term undertaking outlasting the development and construction phases of work offshore and onshore. It is an activity which uses a lot of labour and which has to upgrade the skills of locally recruited employees as we will see in Chapter 6: it therefore provides jobs and training. Processing provides few jobs and most of the locally recruited personnel would work in unskilled occupations. If offshore servicing had attracted state aid Aberdeen's loss of full development area status might have favoured Peterhead and offset some of the disadvantages of distance from the railhead and poor roads. Secondly, existing bases might have taken on additional employees and made innovatory developments for the future if they could raise funds or tax relief from the state.

It is certainly anomalous that the most likely sector for economic growth, long-term employment and the stimulation of small-scale engineering and electronic plants (in maintenance, not manufacture) does not qualify for development aid. If the provision of jobs and skills has been a major objective in the northeast then the equation of development with manufacturing jobs is likely to subvert that objective because manufacturing and processing in petro-chemicals are capital intensive, not labour intensive, whilst 'service' activities create jobs, skills and spin-off developments. Service activities would perhaps have a stronger multiplier effect than processing, in pulling more money into the local economy through wages and the creation of extra demand for consumer goods and services. If it is agreed that private economic enterprise should receive state aid in development areas, it seems illogical to exclude the kinds of firms that are likely to operate in Peterhead. Meanwhile planners and councillors think in terms of finding factories for industrial estates or encouraging petro-chemical processing when it would be easier to encourage activity in the 'service' sector.

In the northeast context the notion of 'service sector' is itself anomalous and confusing. Service enterprise is seen as *following* industrial activity, it is a *tertiary* activity dependent upon primary or secondary production. But the term 'service' has more than one meaning; the 'service sector' includes retailing, banking, insurance, hotels, catering and so on; these are services to industry and consumers but not producers of goods in themselves and therefore in the narrowest sense 'non-productive'. The servicing activities in Peterhead do not produce goods either but neither are they services in the sense of non-productive. If oil production is primary production, then offshore supplies and servicing are part of it. They might even be defined as a pre-primary sector. It could be argued that banking is a similar service, but the location and type of bank (or indeed the source of capital and 'banking' facilities) is unimportant from the point of view of facilitating production. To make oil production possible bases can only be situated in certain locations, and onshore bases are an integral part of offshore work. If oil companies were also base operators, it seems probable that the costs of running a base would be tax allowable as a cost of production. That they are operated by servicing companies as such puts them in a non-manufacturing, non-processing sector and therefore outside the category of enterprises qualifying for state aid — even though, paradoxically, the state had to go into business to promote offshore servicing itself. The use of categories like 'servicing industries' or 'tertiary sector' is relatively

meaningless where one company may encompass a wide range of production and (traditionally) service activities or another may provide
only a service without which production is not possible. That development aid policies are none the less based upon such distinctions makes
them singularly inappropriate to Peterhead.

In choosing examples of the way in which national policies fail to
match the needs of Peterhead, I am not suggesting a conspiracy against
Peterhead. The state develops policies to cope with what is defined as
a regional problem and it is within the limits of such policies that
regional and local governments have to deal with particular local con-
tingencies. The problem of Peterhead would not make the state alter
national policies and in this sense Peterhead is a peripheral location
subject to policies devised to cope with problems elsewhere. The wry
observation was made by a number of informants in Peterhead that had
oil been struck in the English Channel there would have been no wage
restraint. In other words the interests who would have experienced
'unfair competition' for labour in the southeast would have had sufficient political influence to change the course of incomes policies.

We have undermined any notion of Peterhead as an autonomous
community. The town is relatively unable even to profit from its newly
developed assets because of the way in which the locality is dependent
upon and subject to the effects of policies formulated by the government. New developments do not generate additional local rating income,
wealth generated in the locality does not immediately contribute to the
solution of local problems. Furthermore existing local assets, most
notably the harbours, were threatened by proposed developments when
the state (in the form of the Scottish Office) was backing the developers
because of the equation of the developers' interests with those of the
nation.

Planning policy

Without the constraints of law and rational administration the oil-
affected areas would have been taken over by the highest bidders, and
firms like Cromarty Firth Developments or Peterhead and Fraserburgh
Estates would have been in control of substantial areas. Through their
control of land for industrial and housing use these private companies
would at the most basic level have been able to control labour migration
and population change. Their powers would have been equivalent to

those of a small state. But the state intervenes to regulate and prohibit such eventualities. Ross and Cromarty County Council was able to assert the primacy of its own plans over those of CFD. In none of the districts we studied was the state able to protect local authorities from the effects of speculation on the *price* of land.

There is some control over private capital afforded by the procedures established and the relatively powerless individual or small town is not entirely at the mercy of the speculator or large corporation. In many third-world countries, by contrast, tens of thousands of population may be forced to move to make room for capital-intensive agricultural or industrial 'development'. The government and officials may be in collusion with and have a direct financial interest in the incoming corporations. In the UK the existence of, for example, planning procedure does not remove conflicts of interest but it changes the form of their expression. The conflict is an ordered and relatively open conflict conducted according to rules and there has to be a measure of compromise and an adjustment of interests theoretically in accordance with legal, rational and scientific criteria. A compromise entails some interests being over-ridden and in the case of Peterhead it is unlikely that local objections would have stopped a project of 'national' importance, and few living outside Peterhead would think that this should be otherwise. The planning process ensured that Shell/Esso's proposals received maximum public scrutiny and that their assumptions were tested openly. The outcome was a defeat for the applicants, and the final analysis of their harbour proposals showed that Peterhead might have lost its all-weather fisheries harbours had the proposal gone ahead unchecked. Peterhead Town Council was not a principal in either of the two major enquiries. It was abolished between Scanitro's application and the hearing. Thus whilst consultation took place, the town was not formally represented on matters vitally affecting it. It could not be said that Peterhead had participated in decision-making.

The way in which conflict is organised is of some importance to our analysis because planning decisions derive from a mixture of technical and political considerations. A developer has to show that his plant will stand up and not blow up, but beyond this he has to negotiate the conditions under which planning permission will be given. In gaining permission and favourable terms it helps the developer to have influential people supporting them and favourable coverage by the media. It is also valuable to have access to local knowledge and informal access to decision-makers or government. This is why the ex-HIDB personnel were important to CFD. Similarly, incomers to Peterhead sought the help of Jock Smith and he, in turn, claimed to have access to key personnel in the Scottish Office which was pro-development. PFE had as consultant architects a partnership which included the recently resigned Director of Planning for Aberdeen County. Arunta hired Ian Craig, a member of the Town Council and Harbour Trustees, who was also on the county planning committee. PFE, Site Preparations and others sought to wine and dine town councillors, Harbour Trustees and public officials — with varying but apparently slight degrees of success. None of this is to suggest that any of these individuals in fact helped the developers in any special way at all, only that such personnel are felt to be useful by the developers.

We have dwelt for some time upon 'the state' in stressing the dependence of a town like Peterhead. It would be a mistake to see the state as monolithic and omnicompetent. Just as the Shell/Esso planning enquiry showed the corporations to be inefficient so also are there examples of conflicts within and between state agencies and between levels of government which have had consequences for Peterhead. This is clearly seen in the failure of the Secretary of State to develop 'an oil and gas strategy for Scotland . . . and to indicate the role of the Buchan area in that strategy'.

The Secretary of State is in a similar position to a local authority — he needs to develop resources and technical skills for research and the evaluation of policy working from the same poor data base in the same conditions of uncertainty. There is uncertainty about the technical means of recovering oil resources, disagreement over the magnitude of profitably exploitable resources and arguments over the effects of taxation on profitability. The oil industry does not proffer neutral advice on these questions and yet the Secretary of State is largely dependent on the oil industry for information. The industry is therefore in a strong bargaining position and especially so given that they are the main agents for realising the government's economic goals. Thus, for example, ' . . . recent predictions of delay or even a rundown in North Sea Developments, resulting from the tax and participation proposals, amount largely to a negotiating posture by the oil companies'.*

The establishment of BNOC was intended to reduce this dependency

*J. Francis and M. Swan, *Scottish Oil Shakedown*, Church of Scotland Home Board, 1975.

of the state upon the companies while tying its interests more closely
to theirs.

Policy decisions by the Scottish Office directly concerned with
Peterhead related to the harbour and supply bases. Both were to be
developed in the national interest and with little reference to local
interests. It was assumed that the locality would be able to adapt advan-
tageously to these projects.

The Bay Management Company manages the Secretary of State's
harbour and advises him. It is meant to operate as a commercial undertaking but it is not allowed to raise money in the open market. Its
degree of independence from the Scottish Office is not clearly defined
although ultimately it is advisory and subordinate. The company's
freedom of action seems variable and negotiable; for example, the
chairman sought to reassure Scanitro during the Shell/Esso enquiry and
was invited to talk to Scanitro management. The responsible official
at the Scottish Office tried to censure him for acting without authority,
but he withdrew when the Chairman offered to discuss the matter with
the Secretary of State.

In some respects the company operates just like a government department; it issues bills for dues in the harbour but is not responsible for
collecting the money on the principle that no government department
both charges and collects. Whilst this may be an excellent arrangement
for taxing the nation, it creates difficulties for the manager of a busy
harbour serving traffic that turns around very quickly. Thus the officials
and board are not entirely clear about the scope of their discretion to
take initiatives and yet not subject to detailed policy guidance from
the Scottish Office.

These examples illustrate how even when the point has been made
that what happens in Peterhead is largely dependent on external in-
fluences, and that an important influence is the state, the state itself
is not a single, smoothly operating organisation administered according
to rational and clearly formulated and unequivocal policies. Discussion
of the state often assumes, for the sake of simplicity, that it is homogeneous, and then turns upon considerations of the extent to which
the state represents class interests or the way in which it relates to
multi-national and trans-national corporations. This high level of analysis
is extremely important but it is a different and equally important sociological task to analyse the internal political and administrative operations of the various departments of the state. Our main interest is in
the way in which the outcome of these affects events in Peterhead. The

outcomes of encounters are by no means foregone but to a degree
negotiable, although in the final analysis the town of Peterhead itself
lacked any autonomous power.

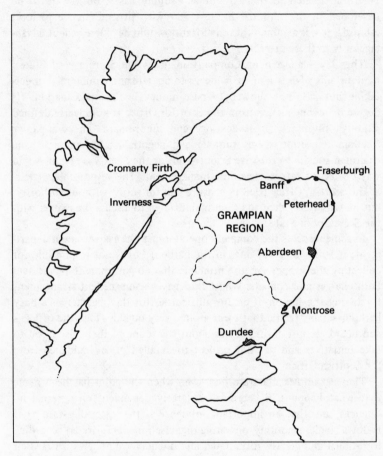

Map 5   Grampian Region

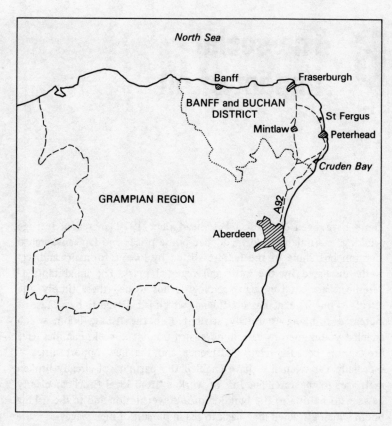

Map 6   Banff and Buchan District

6 The social impact of oil

There have been changes in Peterhead since 1970; increased teenage drinking, marital breakdown, higher house prices and larger wages, to mention but some of the features listed by local informants and regularly discussed by the local and regional press. The inhabitants of Peterhead are good enough sociologists not to call these simply 'the effects of oil'. The increased affluence which is believed to be a cause of increased drinking is usually attributed to the fishing boom which enabled young men to earn hundreds of pounds a week. Families have broken up because relative affluence and the new opportunities — especially for women — have enabled the partners of already broken marriages to separate. The loss of workers from local firms can clearly be seen to be due to the building of a power station and to the fishing boom which attracted men back to sea in pursuit of high wages.

Some of the effects of oil have been indirect, and possibly hidden. For example a fitter and turner may leave an engineering works where he is a machine operator to work on a construction site. His ex-employer may then recruit a man from a small workshop who is, in turn, replaced by the sole employee of a garage proprietor who loses business and eventually ceases all repair work. In an extreme case the garage may close, leaving the proprietor, a pump attendant and an odd-job boy out of work. This chain of events is an effect of oil although the reason why it is more difficult to have a car serviced is not because a garage mechanic has 'gone into oil'.

We need therefore to exercise considerable caution in interpreting the gross changes that have taken place; it may be the case that the most significant changes are indirect, long-term and at present apparently slight. One hypothetical example would be the closing of the garage and the consequential loss of jobs described in the last paragraph. The most obvious changes that have taken place can be described under the following headings: employment, housing, amenities, 'social problems', entrepreneurial activity, unionisation, affluence and poverty.

Employment

The single most important benefit that oil would bring to Peterhead, according to the supporters of local oil developments, was jobs. Peterhead and the Buchan area suffered from high unemployment and out-migration in the 1960s. Unemployment reached peaks of three times the national average and the surrounding district had lost over 15 per cent of its population between 1951 and 1968 whilst Peterhead gained 3 per cent in the same period. The opportunities offered by oil were seen largely and simply in terms of work. This was the basis of the stand taken by the Buchan Trades Council on the questions of Scanitro and the Shell/Esso NGL plant planning applications. Jobs meant more than simply an end to unemployment but also opportunities for young people to stay in the vicinity and an end to the monopoly of local employers whose position in the area resulted in wage levels of 75 per cent of the national average in the 1960s. The local employers, for their part, were, as we have seen, very worried by the possible loss of workers to the oil industry and the pressure on wages.

The total number of employed persons in the Peterhead area rose from 7,567 in 1971 to 10,026 in 1976, an increase of 33 per cent. During this time the number of registered male unemployed fell from 339 to 246 or from 4.2 per cent to 2.4 per cent (2.1 per cent if one excludes 27 school leavers from the latter figure). Plainly the additional 2,500 workers were not drawn from amongst the local unemployed. The major factor contributing to this rise in the number of employed was the build-up of construction work at Boddam and St Fergus (and to a smaller extent on local housing sites). This build-up is shown in Table 4.

TABLE 4

	Dec. 74	June 75	Dec. 75	June 76	Dec. 76	Aug. 77
St Fergus	310	490	1445	915	920	75
Boddam	229	722	819	1120	1251	1442
Total	539	1212	2264	2035	2171	1517

The total build-up is shown in Figure 4.

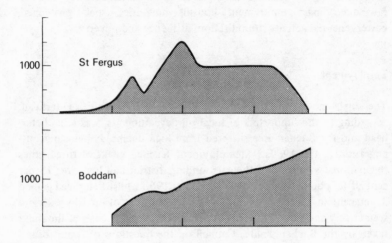

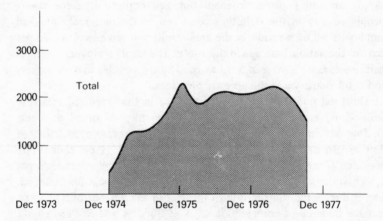

Figure 4   Construction labour; St Fergus and Boddam

The Boddam workforce was not studied because it lay beyond my capacity in the time available. Such are the exigencies of research funding that at St Fergus the work begun at the Total/BGC site was near completion and the grant finished before we were able to take advantage of the contractors' permission to study their labour records. The records of two civil engineering contractors were, however, analysed and we will refer to them as firms X and Y respectively.

Firm X had employed or was currently employing a total of 547 persons: 64 (12 per cent) had been recruited from Peterhead, 83 (15 per cent) from Fraserburgh and 202 (37 per cent) from Banff and Buchan. Thus 64 per cent of the workforce ever employed was recruited locally. Over half (57 per cent) of these had previously worked as general labourers. Of those who had left the firm by the date we collected the data (464 hands) 50 per cent had worked for the firm for 3 months or less. The most common reason for leaving (49 per cent of cases) was absence and poor time-keeping.

Firm Y had a workforce of 107 on site and no records of past employees. 39 (36 per cent) had been recruited from the Glasgow area and 23 (21 per cent) from St Fergus — probably from firm X. Only 3 per cent were recruited from Peterhead and 13 per cent from the rest of Banff and Buchan (including Fraserburgh).

These figures demonstrate quite simply what we observed in Peterhead and throughout the oil-affected areas of Scotland: recruiting policy varies widely from firm to firm. But where did locally recruited workers come from? Of those recruited by Firm Y within the Grampian Region the largest single previous occupation was 'general labourer', followed by housewives and then a wide range including fitters, cleaners, farm labourers, machine knitter, heavy goods driver, and others. Only two were previously unemployed. Incoming firms seemed reluctant to employ the unemployed on the grounds that there must have been a good reason for their being unemployed.

Two points may be made therefore; in the order of *1,000* jobs were created in the construction industry but these jobs were likely to be of a relatively short-term kind. Secondly, substantial numbers of workers were brought in from outside the area. The large and temporary nature of the construction labour force created problems for our analysis because its absolutely large numbers mask significant changes in that part of the labour force which is less temporary. These changes may be of significance for the longer-term future of Peterhead.

The indigenous construction industry was small, and many building craftsmen worked in other trades — the prison, for example, as we will see below. But in 1972 the local construction labour force reached a peak and if we take this as a 'true' local figure for construction workers (after a long depression in building) and discount all other construction workers altogether we find the adjusted total to be 7 per cent of the

Peterhead labour force, which approximates to the national average for construction labour. This adjusted figure is used below and it gives us an adjusted increase in employment from 7,567 in 1971 to 8,394 in 1976, an 11 per cent increase.

The industrial distribution of the workforce is shown in Table 5 and the percentages in Table 6. The category 'Agriculture and fishing' under-represents fishing because the inshore fleet is largely made up of share-fishers who as self-employed persons do not appear in the employment statistics. From more detailed, unpublished, statistics it is clear, however, that agricultural employment is continuing to decline. Some trends are clear from Tables 5 and 6.

TABLE 5   Peterhead, employees by industrial sector

	1971	1972	1973	1974	1975	1976
Agriculture and fishing	794	792	734	686	618	596
Manufacturing	2,957	3,195	3,443	3,278	2,940	3,010
Construction	593	581	581	581	581	581
Services	3,277	3,321	3,486	3,695	3,943	4,102
TOTAL	7,621	7,889	8,244	8,240	8,082	8,289

TABLE 6   Peterhead, employees percentage distribution by industrial sector

	1971	1972	1973	1974	1975	1976
Agriculture and fishing	10.4	10.0	8.9	8.3	7.6	7.2
Manufacturing	38.8	40.5	41.8	39.8	36.4	36.3
Construction	7.8	7.4	7.0	7.0	7.2	7.0
Services	43.0	42.1	42.3	44.8	48.8	49.5
TOTAL	100.0	100.0	100.0	100.0	100.0	100.0

Manufacturing, including primary extractive activities, is declining steadily as a proportion of the whole and the service sector is increasing. But what are the meanings of these trends: is Peterhead's industrial employment declining as non-productive activities expand? A comparison with Scotland and the UK will enable us to put this question in perspective. Figure 5 shows that Peterhead had more of its workforce in the manufacturing sector than either Scotland or the UK. The three largest employers in the town account for this almost alone. The inclusion of share-fishers in the workforce might depress this percentage to nearer the UK average. But as share-fishers are a very small percentage

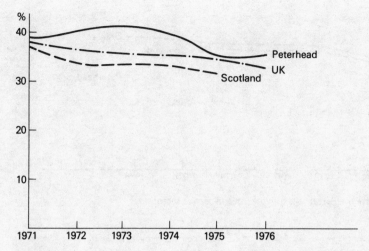

Figure 5   Percentage of workforce in manufacturing, UK, Scotland and
Peterhead

of the total workforce they might be regarded as a bonus, an addition to
the 'productive' workforce in Peterhead rather than simply a factor
depressing the percentage in manufacturing.

The dip in manufacturing in Peterhead in 1975 is accounted for by a
decline in sections of the local food production industries which, in fact,
restored the 1971 level of employment after an exceptional peak in
1973. Overall Peterhead showed about the same *rate* of decline in manufacturing as Scotland and the UK, but it started with a slightly larger
manufacturing sector and so maintained its above average level of
manufacturing employment.

Figure 6 shows another distinctive feature of Peterhead: its small
service sector compared with the UK and Scotland. In professional and
scientific employment, for example, Peterhead had 9 per cent as against
13 per cent for the UK in 1971 and 10 per cent against 16 per cent in
1976 — a smaller sector growing more slowly than the national professional and scientific sector. This is probably a reflection of Peterhead's
peripheral status: planning, research and development, administration
in both private enterprise and public services, are conducted in Aberdeen or other city centres. But the whole service sector in Peterhead,
although 9 per cent below the national service sector, is increasing
sharply. Both sectors for Peterhead, the UK and Scotland are shown in
Figure 7.

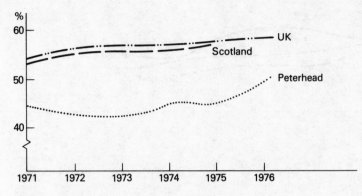

Figure 6   Percentage of workforce in service sector

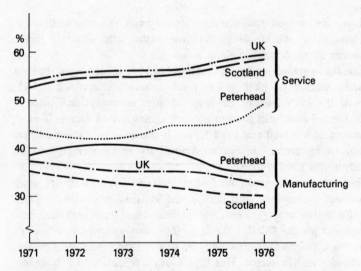

Figure 7   Workforce in manufacturing and service sectors, UK, Scotland and
Peterhead

When we examine the increase in employment in terms of jobs for men and women we find that women's employment has expanded from 2,762 to 3,598, an increase of 30 per cent, whereas male employment has fallen from 4,805 to 4,796 (or, in practical terms, it has remained steady). Yet, as we will see below, a number of firms reported a loss of male employees during the period, and especially in 1973; where did these men go? Some may have returned to fishing or they may have moved into that part of the construction industry which we have defined as temporary and therefore excluded from the statistics. If local men had moved into construction then we might have expected a rise in unemployment when construction declined, and this was the case in late 1977 and early 1978.

The most dramatic increase has been in the number of women employed in the service sector. The growth has mainly been in postal and telecommunications services, retailing, banking, hotels and catering, educational, medical and national government service. Together these account for 71 per cent of the increase in female service sector employment for 1971-6, with hotels and catering showing the largest single growth by nearly doubling. In manufacturing there were 408 extra jobs for women in 1973 as a result of a temporary expansion in food processing. The main decline in male employment was 25 per cent in agriculture which represented the continued 'push' from agriculture by mechanisation and the consolidation of farms and the 'pull' of high wages on a poorly paid industry. There was a small net loss of male manufacturing jobs — less than 1 per cent of the 1971 male workforce. The male and female workforces are compared in Figure 8; in 1971 37 per cent of Peterhead's workforce was female compared with 38 per cent in the UK, in 1976 43 per cent was female compared with 41 per cent in the UK.

In summary of these general trends we may say that in the period 1971-6 Peterhead saw no major loss of manufacturing employment but increased employment in services, a very substantial proportion of which was taken by women workers who at the end of the period were a larger proportion of the Peterhead workforce than of the national workforce. There was also a very large increase in construction work very little of which is likely to be permanent.

TABLE 7   Peterhead, employed men and women by industrial sector

	1971		1972		1973		1974		1975		1976		1971-6	
	M	F	M	F	M	F	M	F	M	F	M	F	M	F
Agriculture and fishing	715	79	724	68	656	78	616	70	535	63	531	65	−184	+14
Manufacturing	1,910	1,047	1,854	1,341	1,988	1,455	1,891	1,392	1,872	1,069	1,877	1,137	−33	+90
Construction*	508	31	547	34	547	40	547	53	547	81	541	141	+33	+110
Services	1,672	1,605	1,645	1,675	1,707	1,776	1,691	2,004	1,865	2,078	1,847	2,255	+175	+650
Total	4,805	2,762	4,770	3,118	4,898	3,349	4,745	3,519	4,819	3,291	4,796	3,598	−9	+864

*Corrected to allow for migrant construction workers

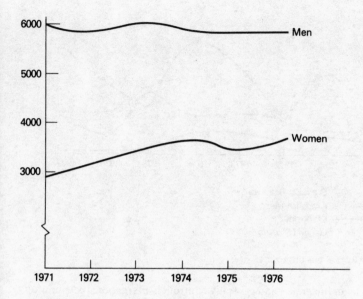

Figure 8   Men and women in the Peterhead workforce

Effects on specific employers

In 1970 the four largest employers in Peterhead employed about 1,100 persons: the prison employed a further substantial workforce in excess of 150 persons. All these employers were to feel the impact of local economic changes. The simplest general measure of the effects is labour turnover, and we obtained data from three firms which enabled us to calculate their labour turnover. A fourth firm supplied the turnover figures which they themselves had calculated. Labour turnover for the four firms is shown in Figure 9. The presentation of even the most elementary details of these firms would make them immediately identifiable and there is thus little point in observing the usual convention adopted by sociologists of disguising the identity of their subjects. *General Motors* employed 475 workers in the production of transmissions for the Terex earthmoving equipment division of the company. The Peterhead plant is independent in terms of day-to-day management and operates within clearly defined policies and controls exercised on an international basis. The plant supplies the Lanarkshire works with gearboxes for earthmoving machinery. In other words, it is a branch

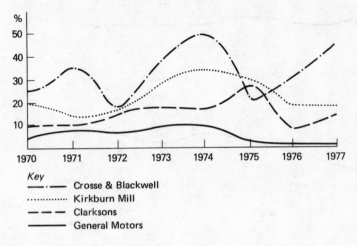

Figure 9   Labour turnover 1970–7

plant in an integrated and centrally controlled international corporation. The workforce is trained on the job with initial off-the-job training, and General Motors is reckoned to be the local leader in wages and terms and conditions of employment. These conditions are influenced by local policy integrated with the policy of the parent company and international policies of the corporation. According to management they had a waiting list for jobs of 200–250 which included people who already had jobs elsewhere. When General Motors took over the previous firm on the site in 1956, wages were 25 per cent lower than those paid in the Lanarkshire works of General Motors, but this gap gradually closed and in 1969 it became policy to eliminate what remained of it. The difference was reduced to 4 per cent before statutory wage restraint prevented its being reduced further.

Labour turnover in the 1960s was between 3 and 4 per cent and was at 4 per cent in 1970. The firm had been accustomed to higher turnovers in fishing booms and one such peak occurred in 1971, but they had adapted to this problem by limiting their recruitment of ex-fishermen. The very substantial rise in turnover in 1973 and 1974 coincided with the build-up of construction activity at Boddam and St Fergus and a small growth in employment at the bases. This, according to a manager, 'gave management a shock'. The high rate of turnover declined quickly and in 1975 was back to 4 per cent. By 1977 ex-employees were returning from oil-related jobs. It seems that in 1973–4 there had been a

once-and-for-all outflow of men from General Motors, comprising those
who thought construction would offer 'big money' and building workers
who had taken jobs with General Motors because there were no opportunities to practise their trade.

By the time this research took place the local management of General
Motors had recovered from the shock of the peak turnover in 1973 and
1974. Had this been a trend, then General Motors might have found
itself unable to train men fast enough to keep up with labour losses.
The labour question in relation to oil was having an influence on their
thinking about future investment and expansion because it introduced
an element of uncertainty into an otherwise apparently secure future.
Some plans for expansion may have been held up and whilst the division
represented a heavy capital investment and an accumulation of valuable
skills such a small element in the General Motors enterprise could be
moved to an area of the world where the labour supply was more certain
and wages under less local pressure. Increased automation of the work
of the division could de-skill the jobs to make them more transferable
and cheaper. The apparently temporary nature of the additional jobs
created in the locality now reduces this risk for Peterhead.

Clarksons Tools belongs to Clarksons International, a wholly owned
subsidiary of Thorn Electrical. The firm makes high-quality drills and
reamers under the 'Cleveland' trademark and has a 'knowhow' agreement
with Cleveland Tools (Ohio) which enables it to draw upon United
States research and development.

None the less it is not simply a branch plant. Peterhead is the headquarters of Clarksons and includes the marketing division, engineering
and design, senior management and nearly 100 clerical and supervisory
staff. Clarksons has branches elsewhere. The products are both to contract and stock, with outlets in Manchester, Birmingham, London and —
through Glasgow — Holland. Special steel for their products is supplied
from Japan and the USA. In no sense therefore is Clarksons a 'local'
firm.

The manual workers number over 400 but only 30 are time-served
men, working on maintenance and tool-making. From 1970 onwards
Clarksons had what a manager described as 'traumatic experiences, it
would be fair to say'. For fifteen years or more before the present management took over in 1971 there had been industrial relations problems
which will be described later in this chapter. In early 1972 and 1975
there were wage disputes, which in 1972 entailed a seven-week district
strike by the AUEW. Also since 1970 there have been periods when

business was at a low ebb with the firm on a four-day week. Between autumn 1974 and autumn 1975 over 150 people left as the business cycle declined. In early 1977, when management were interviewed, the firm was working at two-thirds capacity.

Clarksons have never experienced difficulty in finding workers. Their main problem was the attraction of General Motors which in 1977 were paying £15 a week more than Clarksons and therefore men who had gained experience at Clarksons were likely to move to General Motors if the opportunity arose — even though the larger employers in Peterhead operated a nominal 'no poaching' arrangement.

The effects of oil became obvious in 1974 (when turnover was lower than either 1973 or 1975) and, according to the management, 7 or 8 people were leaving every week, 'including some of the firm's best'. In nearly all cases we found that managers speaking off the cuff exaggerated the turnover; our statistics show that in 1974 Clarksons lost an average of 22 persons per quarter, or less than 2 per week. Construction tradesmen left for construction work in their trades. Skilled men left for St Fergus and others to Boddam and the bases. Others left for labouring jobs and driving earthmoving machinery. According to management, the hourly rates of pay were not better than at Clarksons but the leavers were able to work longer hours. Some left saying 'we'll be back' and by early 1977 many were back, including some who had been contacted directly by management and persuaded to return. Interestingly the management also believed that a number of men had left for store-keeping and record-keeping jobs because these gave men greater autonomy, less supervision and more freedom of movement which, irrespective of wage levels, was an improvement on working in a factory. Oil also brought some benefits in that workers came to Peterhead looking for work in oil 'on spec' and, finding nothing available, came to Clarksons.

The unions contrasted wages and conditions at Clarksons with other employers and also harked back to a time in the 1974–5 period when they worked half-time, half the workforce worked the first half of the week and the other half the second. Workers were averaging £25 which, with benefits, came to about £35 a week. In 1977 'at Clarksons you would have to work two nights and Saturday to get the same wages as General Motors and of course at the various oil-related sites the bonuses are fantastic even though the basic rates of pay might be quite poor. [Another firm] have £40 a week bonus whereas Clarksons pay only one and a third or one and a sixth — there is something like £16 difference between Clarksons and comparable employers'. But the firm should have no difficulty in recruiting workers, even those who leave for oil or fishing are quite likely to come back.

The union leaders were also acutely aware of the status of their firm. Prior to being taken over it was playing a classic role for a firm in a peripheral area — it was sending semi-finished goods to Holland, thus avoiding high tariffs. The goods were then finished in a factory in an area of high industrial unemployment where the wages were heavily subsidised. But the unions saw the future as secure. In this they were not at one with all the management some of whom believed that another branch in Sheffield could take on their work if the conditions of the local labour supply and wages warranted it.

Textiles were once a major industry in the northeast but this has not been the case for nearly two centuries. However, one textile works is to be found in Peterhead — the *Kirkburn Mill*. This mill has been in the Smith family's hands for 150 years. It is equipped to carry out almost any task and can receive a fleece and despatch either yarn or cloth. The mill is plainly very vulnerable to the loss of key workers because without certain skills being available the mill will stop. Two-thirds of the 200 employees are women but they rely upon men to do certain heavy jobs (in the dye house for example). Most of the men are chargehands and foremen.

Textile producers feel market pressures acutely and from discussions with managers in 1977 it was clear that such issues as foreign imports, overseas aid to third-world competitors and price inflation were issues that taxed them as much as the effects of oil. Although a 'local' firm, unlike General Motors and Clarksons, the Kirkburn Mill is as dependent on New Zealand (for raw materials) and Yorkshire (for markets) as General Motors and Clarksons were upon their parent companies and overseas branches.

By 1971 the worst effects of the oil were, from the management's point of view, over. But the mill was felt to have suffered badly, men left to seek higher basic wages and opportunities for overtime. In one shop, with a foreman plus two men, one man gave notice on the day I first visited the mill and the other, with twenty years' service, was hoping to leave. Meanwhile the foreman was having to do the work of the man leaving. The Employment Services Agency cannot send a replacement because the job requires experienced hands. In the dye house, where the work is least skilled and most unpleasant, the work was being done by men who stayed two months or less and the mill was having to commission dyeing in Yorkshire. None the less the peak of the oil effect was passed by 1977; as Figure 9 shows, the turnover of 33 per cent in 1974 and 30 per cent in 1975 had fallen to just below 20 per cent. The reason for this was said to be that 'men are no longer impressed by oil which demands a seven-day week and twenty-four hours on call' (a reference to work at the bases and offshore).

The management did not paint a rosy picture in saying that it was hard to understand why any men stayed at the mill: 'Those who did not leave are unhappy. Wage legislation has been a disaster and the place is riddled with malcontents. Their discontent is over wages mainly but also fringe benefits, future prospects and present (labour) turnover.' The mill tried taking on ex-prison officers at one time, but the latter were unhappy with the loss of authority entailed in becoming unskilled manual workers.

The mill had, rightly or wrongly, at the time, a bad reputation for wages in the town. I discovered that in 1970 a few workers received less than £10 a week and in 1978 the basic rate was £35 a week, to which has to be added Heavy Duty and other allowances. The rate was above the nationally agreed minimum wage for the industry but should be compared with the national (UK) average take-home pay for male manual workers of £71.50 in 1977.

The mill seemed to experience the *indirect* effects of oil in an acute way. In the words of Michael Smith

> the local wage structure is out of line with engineering and the mill
> suffers — with Crosse & Blackwell and fish processing. 'Oil' is a
> world apart and will pay anything for anything (and its efficiency is
> therefore very low). But we have to get engineering jobs done from
> Yorkshire, with all this capacity on our doorstep. The man who
> repairs rollers gets £1 an hour on textile machines in Yorkshire,
> locally, from £1.75 to £2. Visiting specialists just cannot believe the
> Peterhead wages. And no local firms will tender for jobs at the mill
> when they can get costs plus with the oil industry.

In spite of all these difficulties, however, the mill never closed. But its position was always vulnerable because plans had to be made and altered on a day-to-day basis whilst 'Shetland has more orders than they can cope with — but we can not begin again once we lose work.' To this extent the mill suffered from not being a branch plant of an international combine.

Crosse & Blackwell is a branch plant of a firm which is a subsidiary of Nestlé whose head office is in Switzerland. The firm appears to be well-integrated locally, taking its raw materials (all its raw materials at times) from local farmers on contract. Potatoes, carrots and beef are especially important.

The workforce varied from 450 to 610 in the period of the study, being mainly in the high 500s. This total includes part-timers. About 80 were on the technical and clerical side and the remainder (largely women) on production. There was a high turnover of women but most of those leaving were short-service (according to local management) and recruitment kept pace with leavers. Normal turnover was about 17–18 per cent and this *low* figure is accounted for by the lack of alternative employment for women. 10–15 per cent of the workforce came from the country but by 1977 a 'large number' of recruits were strangers. Seasonal fluctuations were coped with by natural wastage and 30 or more students were taken on in the summer. The workforce is also the most obtrusive in the town: morning, evening and lunchtime it flows on to the main streets in white overalls and caps.

The wages were JIC rates which might be raised by a third with bonuses. In the week before the first interview at the cannery works early in 1977 the average hours worked were 36.7 and the average wage (depressed by part-time earnings) £36.31 (compared with a national average of £43.70 for manual female workers over 18 years of age).

The problem created by oil was the loss of male workers: 'a lot' left to go into construction and technically qualified men left for the bases. Tradesmen were difficult to recruit but the management were able to help recruits with housing and sought men when closures took place in the food industry further south. Skilled, time-served men are needed to maintain the production lines and keep them moving. Less-skilled jobs, like driving fork-lift trucks have been taken over by women – who needed considerable persuasion to take equal pay with the male drivers. The other major problems related to expansion. In the short term two 8-hour shifts were needed to cope with demand for canned foods but only one could be manned because there were not enough women available. In the longer run the firm as a whole had plans for expansion which would have required an extra production line in Peterhead. It was extremely uncertain in 1977 whether production or maintenance workers could be found to make such a development possible.

On the whole management felt that their real difficulties would not outlast the construction phase in the oil industry and that already men were returning and expressing a preference for indoor work. Petro-chemical developments, on the other hand, would bring more workers into the area and they would have wives seeking work. So Crosse & Blackwell stood to benefit. We will need to consider the implications of the growth of services, offering work for women in the area rather than in the petro-chemical and downstream processing. Service activities in the bases and the growth of offshore supply and maintenance enter-prises is likely to sustain the demand for the skilled workers that neither the mill nor Crosse & Blackwell can obtain. The position of Crosse & Blackwell will, however, remain crucial in the local economy and this position is widely recognised. Its fiercest critics concede that any small town would 'give its right arm' to have a firm offering 500 jobs for women.

The only other large local employer (and one of long standing) is *HM Prison, Peterhead*, 'the Dartmoor of Scotland' and as a posting for officers from the Midlands perhaps no less popular than for their charges. The details of employment within the prison service may be an Official Secret but the total number of staff is not far off 170. There has been a recruiting problem for Peterhead prison and staff shortages have been met in part with overtime payments to existing staff. The amiable Chief Clerk Officer had kept a note of every person leaving the staff since the early 1960s and in one respect his records were superior to those of other local employers; he had noted leavers' trades.

Table 8 shows numbers leaving the prison service from 1972 to 1977.

TABLE 8   Officers leaving prison service: HM Prison, Peterhead

Year	1972	1973	1974	1975	1976	1977
No.	44	20	7	5	6	1

Of the 20 leaving in 1973 9 were in construction trades, 5 returned to fishing, 1 to driving and 1 to 'oil'. In 1974 2 returned to fishing, 2 to labouring and 1 was a welder. It would seem that in 1973 the prison was sharing the experience of other employers, losing construction tradesmen who returned to their trades and fishermen who returned to the profitable life at sea.

So far we have discussed the large employers in Peterhead and it was from them that we obtained detailed statistical data. But, like most towns, Peterhead has a large number of very small enterprises, some em-ploying less than five people. Three kinds of employers of medium and small labour forces have been affected by developments in Peterhead: those concerned with construction, banking and hotels and catering. The bankers and hoteliers have had to cope with expansion and the construction trades with considerable pressure upon their existing labour as a result of the construction boom.

The local building industry consisted of a number of small firms loosely organised in a consortium. Thirty-three building tradesmen were listed in the 1977 Peterhead Yellow Pages of the telephone directory. No one builder was 'all trades'. So when building needed to be done one firm took the main contract and sub-contracted to the others. Each firm specialised in a particular trade — joinery, plumbing, slating, etc. — and they undertook work for one another on a credit basis. Land for house-building was purchased in small parcels for about a year's needs and the money from house sales was used to buy more land. The 'organisation' of the building industry into a loose consortium seems to have been the result of local authority pressure. The local authority found it administratively impracticable to be dealing with a multiplicity of small firms in order to have one building job satisfactorily completed.

The local building industry was probably no better or worse organised than any other. It enjoyed a boom with fishing, and houses were built on the south road by the bay and to the north by the River Ugie — where one builder erected about 200 houses. The local industry could not, however, cope with a power station, gas terminals, bases, a new Academy building and with council housing on the scale envisaged in the various plans discussed in Chapters 2 and 3. For these contracts major national firms moved into the area. At the same time — as we have seen — land prices rose so that one builder who had been paying £2,000 an acre for house-building land found he was priced out of the market at £25,000 an acre. He also lost one-third of his skilled tradesmen to the contractors on the major sites.

How, then, did these small local firms cope with the influx of large construction companies? Firm A simply allowed itself to be taken over by a large incoming firm, thus providing the incomer with a local foothold, with expert knowledge and a nucleus of labour and transport. Firm B had 26 tradesmen in 1970 but in 1973 had 12 only and the owner reported that he *could* get labour if he was prepared to pay for it — the men would go to the employer offering the best wages. In mid-1977 he lost 2 tradesmen who had been with him for 20 years. The boss reckoned tradesmen could get over £2 an hour and £1.30 on average. The 2 men he lost were to get £4 an hour as pipe-fitters, wages with which he was not prepared to compete unless he had a job that made it worthwhile. The firm now does small contracts at Boddam and St Fergus but the main expansion has been into ship work. This used to be 15 per cent of B's work but is now 60–70 per cent of the work. One vessel had employed 7 men for 140 hours and they each pocketed £250 for a week's work. B, in common with other builders, picked up some of the finishing work after the builder at Waterside went bankrupt. B had therefore adapted by expanding his ship work and tendering for small but profitable contracts on the major sites. He firmly believed that when the 'big boys' went away he would recover his labour force.

Firm C had taken advantage of the boom in house-building and expanded its labour force to over 90, made up entirely of men from the Banff and Buchan area. The expansion of house-building has been made possible by the introduction of systems building. C had become an all-trades firm but had few bricklayers; the firm was able to expand its complement of joiners, plumbers and electricians to cope with the new-style building. The firm also renewed its equipment and expanded its storage. This expansion was financed, it seems, through a successful housing development outside Peterhead prior to oil. C was also able to build nearly 100 council houses by the use of the new methods. In 1973 they finished 300 houses. The only worry about labour was the ability of General Motors and Clarksons to take their men.

Firm D was mainly concerned with bricklaying and roofing. In 1973 it lost a quarter of a workforce of 40 men. Those who left were tradesmen and were never replaced. The present workforce is 20, half tradesmen and half labourers. D supplied materials both to Site Preparations and to the Waterside site and lost money on both. The Company had survived by gearing its activities down to the available workforce and by diversifying. In cutting down they have refused to take small contracts and thus contributed to a situation described by the boss of firm B, who said it was impossible to get small domestic building and repair jobs done in the Peterhead area.

D diversified into plant hire. This is the most profitable part of the business and part of the labour force has been trained to drive diggers and dumpers. Thus men can work on buildings or as drivers and are continuously in work. D's young manager was careful not to overspend on capital equipment because he felt plant hire would only be needed until the large construction firms established their own bases in the northeast. Some plant-hirers had, in his opinion, overstretched themselves – he, meanwhile, was able to undercut them and now found his service much in demand by incoming companies.

The building industry has survived therefore by a variety of strategies, contracting and diversifying, expanding and adopting new techniques. No local building firm had gone bankrupt, although some like D, had suffered some very bad debts due to others' bankruptcy. The builders might have been the worst sufferers from the labour demand created by the construction firms. Construction workers had enjoyed high wages, but non-oil employers had in turn found it very hard to employ tradesmen. It had become virtually impossible to get minor house repairs done.

The banks have had to cope with a very rapid increase in their business in Peterhead, the 'big three' banks (plus the TSB) have all seen their workforces doubled since 1970. They have had no difficulty at all in recruiting employees with relatively high educational qualifications to secure and prestigious jobs in the banks. There were in 1977 71 bank employees in the town. The banks have also had to expand their premises and in the interests of security to alter their counter layouts. The bankers, perhaps embodying the capitalist spirit in Peterhead, have been eager and willing to respond to new needs: one makes up the wages for incoming firms and supplies the made-up envelopes for every worker on pay-day. Another has provided capital for oil-related expansion of local undertakings. One opened a sub-office in an offshore base. All now have to handle large quantities of money, a great deal of foreign exchange and cheques drawn on American banks.

The banks are not local firms and plainly no town the size of Peterhead would be able to support four. They are all branches of national banks and the scope for lending money and taking decisions in the field of business and industrial investment is severely limited at the branch level. But the banks are significant employers, offering some of the few jobs available in the 'white-collar' sector, with training and career opportunities for their employees. The increased numbers would have been very much larger but for the introduction of centralised computerised accounting and the adoption of micro-fiche recording. Given the increased flow of money in Peterhead, based on fishing, and oil, it seems unlikely that there will be any redundancies in the banking field, but if there was a contraction in future staff might be transferred away.

Much of the increased cash flowing through the banks originates from the bars and hotels of Peterhead. It seemed unnecessary to use limited research resources to pursue the entirely obvious fact that the licensed trade in Peterhead was booming. Peterhead's thirteen pubs and clubs employed 62 people in 1970 and 122 in 1976. The town is relatively under-provided with hotel accommodation, especially of the kind thought suitable for company executives. Employment in this sector rose from 140 to over 200 in the same period – most of the increase being for women.

In this discussion of the indigenous industries of Peterhead we have seen the effects on the four largest employers upon whom, in default of petro-chemicals, the long-term prosperity of Peterhead may depend. We have seen that they have all survived periods of relatively high labour turnover. This high turnover was a once-and-for-all phenomenon consisting of tradesmen returning to their trades and seeking high wages in construction. The building industry, small by contrast and consisting of numerous local employers, lost labour and had to adapt either by expanding to capitalise upon the boom or by concentrating on industrial repair, fabrication or servicing that the big incomers preferred not to deal with. I found no job-destruction, although I did not look in detail at smaller firms, like garages and workshops. Of the thirty firms who closed between 1974 and 1978 none could be attributed to oil. The reasons for closure ranged from death or retirement of owners to fire and redevelopment. A number of shops also closed as a result of the parent of their chain deciding to close small branches. Only one case of job-destruction was discovered in a small fish-processing works where so many women left to work in the camps that their jobs were taken by men sent by the ESA. According to the manager the new men got drunk on their first pay-day and left the firm with their week's money, only appearing again to vandalise the building at the weekend. The fish-filleting tasks to which they had been assigned were then transferred to Fraserburgh. In the service sector the banks expanded to cope with increased business but not by as much as they would have done without radical technical innovation. Qualified labour power has been readily available and the banks seem to have obtained their existing labour force because of the security and prestige their employment offers. In hotels, pubs and clubs women (mainly) have been recruited and the nature of the work makes it possible for women with families to serve appropriate hours.

Finally, in this section we turn again briefly to the incoming employers. For the purpose of discussing changes in the Peterhead occupational structure we discounted the large number of construction workers connected with Boddam, St Fergus and the building sites because they

were temporary. Locals were, however, recruited to these sites as we
have seen in the accounts given by local employers, by the local workers
themselves and from the records of firms X and Y. The build-up of the
workforces is shown in Figure 10 and Table 9.

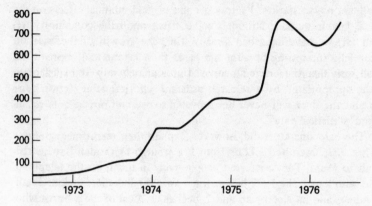

Figure 10   Labour force: bases & clients (not including supply boat operators)

TABLE 9

	June 74	Dec. 74	June 75	Dec. 75	June 76	Dec. 76	Aug. 77
St Fergus:							
construction	15	135	490	1,455	915	920	75
pipe-laying		690	100	400	400	400	
Total	15	825	590	1,855	1,315	1,320	75
Harbours:							
bases	60	200	175	225	295	295	375
clients	50	55	135	160	455	335	320
supply boat ops	50	165	250	660	655	735	500
Total	160	420	560	1,045	1,405	1,365	1,195
Boddam			722	819	1,120	1,251	1,442
Grand total	175	1,245	1,872	3,719	3,840	3,936	2,712

The increase of 2,600 jobs over 40 months, with a peak of nearly
4,000 is, as can be seen, made up of a number of elements: (i) a steady
build-up and decline of the St Fergus Total/BGC site construction force;
(ii) the arrival and departure of a wholly migrant pipe-laying team

encamped in caravans; (iii) the slow but steady growth of the bases' staffs; (iv) the development of the bases' clients' workforce; (v) the growth of offshore activities (both (iv) and (v) grow and decline with the development and completion of work in the Thistle, Piper and Claymore fields); (vi) the growth of the construction workforce at Boddam power station. Factors (ii) and (v) had minimal effects on the local labour market, although some effect upon the community. (i) and (vi) we have described already, (iii), the growth of the bases, is especially interesting because the bases train labour and improve the skills of locally recruited people. Unfortunately my statistical data were appropriated by a research assistant who left after six weeks or so and the data will never be recovered so we will not be able to see 'hard' statistical data.

The base operators did, however, explain their recruitment policies to us. *BOC* recruited labour from Fraserburgh to Cruden Bay and inland to Maud. They preferred people living in town, on the telephone and with a car. The 1977 labour force was 200, but these included only 2 joiners and an electrician and 2 mechanics. Most of the jobs could be learnt on the site, as could safety training. The goods handling, which comprised the major part of work for the manual worker, was mechanised and so 20 men were trained in fork-lift truck driving, 20 as goods handlers, and all were taught how to sling loads. There was a training manager on the site and employees might be trained in all the skills mentioned and also for HGV licences. An effort was also made to predict and train for further skills. It cost BOC about £500 to train a worker in 1977 and it was therefore important to keep their trained men.

The company also provided training for management and administration staff; dealing with such questions as social relations at work, business problems, accountancy and report writing. The intention was to make managers and supervisors multi-skilled and employable anywhere within BOC.

There were 70 female workers on the administration, secretarial and clerical side and 30 cleaners. A relatively high proportion of the workforce was management, employed not so much to manage the manual workers as to manage the affairs of the clients and to co-ordinate the complex logistical and administrative problems of ensuring a reliable offshore supply service. Five of the senior management were recruited from the north and northeast of Scotland.

A few skilled men were recruited from General Motors and Clarksons. At one time the company recruited local fishermen but dropped this practice when they discovered a high labour turnover amongst fishermen.

BOC planned a major £2 million development adjacent to the Upperton Farm site. This was to be spread over two years. The first phases entailed an expansion of warehousing and would create a further 20–30 semi-skilled jobs for which workers could be trained on site. The later phase involved the building of a 50,000 sq. feet fabrication shop employing anything from 50 to 100 time-served skilled workers. Labour recruitment was recognised as a problem in this latter case and BOC needed access to a skilled labour pool without driving up local labour costs.

The ASCo base employed 125 people in 1977. About 115 were local although some of these were originally incomers to the area or returned migrants. They also recruited from General Motors. Locally recruited workers had a bad time-keeping record and the manager felt that although they were willing workers they were not so physically fit as men from further south. They picked up odd-jobbers, seasonal and fishing workers when they first came but later recruited more carefully. At least this was the management account. One of the workers put it more bluntly 'they took on the dregs and unemployables from Peterhead and then had to clear them all out'.

The men worked seven days a week, 40 hours basic plus 24 hours overtime — in the slack season. The firm was unionised by the Transport and General Workers Union and the union worked out the overtime rota.

The bases by the nature of their operation required a workforce that was available at all times, adaptable, trainable and completely reliable. A high proportion of supervisors was also required. This suggests that they either (a) 'creamed' the local workforce or (b) upgraded the skill levels of indigenous workers. My evidence so far suggests that they did both.

The client firms varied in their size and their activities. Some were 'a girl and a telephone'; other quite extensive operations employing hundreds. Some of these clients welcomed my research efforts and gave time to talk to me; others were not contacted due to pressure of time. One of the largest operators in Peterhead promised me access to material but succeeded in finding difficulties in doing this, or had to pass the request around various offices, and I found people 'out' whom I knew to be in. They finally left with expressions of regret at not helping me.

Had they been too busy to help me they would have said so. In retrospect it seems they were trying to refuse without saying so. In the course of my relationship with them, it became known that the company was being investigated for tax irregularities and this may explain their eagerness to keep me at arm's length.

In the end I met managers of four firms operating offshore. The bustle and constant interruptions made interviews difficult, one could only observe and interject the occasional question. The managers were clearly a highly mobile elite, moving between companies and locations. They were almost wholly concerned with activities offshore and left most practical day-to-day matters to the base operators. Firm 1 had 21 personnel, 11 of these were managers or supervisors. The remainder were locally recruited typists and labourers. They used to employ local fishermen but no longer did so. Storemen and men with any skills superior to labourer were recruited by the parent firm and not locally. The firm was only in Peterhead for 18 months. Firm 2 was a management firm, co-ordinating the activities of other operators. The management team was 23 strong. Five typists and a computer operator were locally recruited. This total was thought likely to rise to 8 with the expansion of the firm. Firm 3 was an oil company. The staff consisted of 3 managers, 4 girls — 2 concerned with custom and 2 telex operators, 2 warehousemen and 3 despatchers. Nearly all those staff were poached (in the manager's words) from the base operator and other clients. When the field went into production there were a further 16 persons involved in stores and maintenance.

Firm 4 was another oil company employing 16 people, their heavy labour being done by the base operator. Supervising the base's chargehands, they had 3 foremen, who were recruited locally through newspaper advertisements. The 9 men in the warehouses all needed drilling experience in order to do their job and they were difficult to recruit when there was plenty of work offshore, but 3 were recruited locally. The company had a list of names of suitable people and was prepared to move people to Peterhead at the firm's expense. The recruits had to have been derrickmen or pump men, not roustabouts. Four girls were employed as secretaries. But in recruiting locally they found they picked up men with drinking problems so they started using an Aberdeen employment agency. For secretaries (who were not alleged to have drink problems) the local agency (see below) was used. It seems that 4 was trying to recruit labour at a time when unemployment was at a minimum in Peterhead and only a few remaining unemployables were available through the ESA. Although fishermen were recognised as having a high turnover, the local labour was thought to be satisfactory by most employers.

What we found, then, was a diverse pattern of employment varying from immigration of staff to local recruiting and training. The needs of the client companies and their lengths of stay in Peterhead varied considerably. The base operators had locally recruited and trained personnel and their clients were not beyond poaching staff from them. This also represents the difference between the base operators who have a permanent commitment to the area, wish to maintain a stable local labour force and not to inflate wages and the short-stay offshore companies who, working in an international context of goals and cash, only wish to get a job done quickly and at any price. One or two of the client firms reported 'organised chaos' in the bases as operators competed with one another and within their own organisations. They stressed the brevity of their stay, the continual changing of day-to-day arrangements, communication failures and waste. One employee commented that he was daily amazed that petrol remained so cheap after what he saw at work. The management firms and the base operators had to make some order out of confusion.

The bases themselves offered long-term employment prospects as evidenced by the proposed growth of BOC. The jobs would last for as long as the North Sea remains a profitable source of oil and gas — 25 or perhaps 50 years.

Lastly, we turn to the incoming construction companies. The study was confined to the activities upon the Meethill road, where about 800 houses were being built for the SSHA and the District Council (including some for coastguards and prison). The other major construction job was not oil-related, namely the building of the new Academy. Excluded from our enquiries were the private building schemes involving about 70 houses.

The peak demand for labour at Meethill and the Academy was for about 400 men from mid-1976 to 1978. The Academy site was situated away from other building activity near the town centre. The main contractor employed about 80 men and the subcontractors 30 at the peak in summer 1976. Conditions of employment varied widely between the sites and on one site the senior supervisors were extremely critical of the payment system, which they felt encouraged bad workmanship.

The largest housing site employed over 200 men and was experiencing considerable difficulty in recruiting labour in mid-1977 and their turnover was so high they ceased to record it. They very much wished to recruit local tradesmen, whom they found to be excellent workers, but they had only 4 on the site. Their failure to recruit craftsmen was due to the relatively low hourly rate of pay offered. They sent buses across the Buchan area to bring in men from as far as twenty miles away. They also had 22 chalets on site and 15 caravans for travelling men. The locally recruited labour was unsatisfactory from management's point of view because they would not work overtime. Management also thought they had detected a tendency for locally hired labourers to work for 13 weeks, which would keep them on full unemployment benefit for the following year. Without reference to labour records, such suggestions cannot be tested. The site manager said Peterhead was the worst site he had ever worked on — a view endorsed by the other management staff who crowded into his office to join in our discussion.

The second site employed over 100 men in July 1977. Very few local men were employed, but a bus brought men from Aberdeen and Ellon. Thirty workers came from Aberdeen and the remainder from the surrounding area. About 20 bricklayers and 16 joiners were employed. On the third site the peak labour force of nearly 90 in December 1975 had declined to 73 by December 1977. The site foreman contrasted his workforce with the previous one. He had hired joiners, bricklayers, plumbers, slaters and labourers locally. Others came from Turriff and painters and electricians from Dundee and Stirling. The company paid good wages and bonuses and was well ahead with its work.

The fourth firm recruited in Manchester when it began operations in Peterhead, but by mid-1977 was recruiting locally. They employed 71 men, all local except 2 Irishmen, a Fife bricklayer and 4 painters from Newcastle. At the end of 1977 the company went into liquidation.

I did not attempt to research the labour records of the builders, nor did I speak with many of the men. It seemed simply to be the case that the builders got the labour they paid for. The poorest payer found the local unemployables at his gate and was likely to have to exclude them from the site if they made their way back there after lunch on the first pay-day. The best payers had no trouble finding workers.

One feature of the payment system had bizarre results. If workers lived 30 miles from work they were entitled to tax-free subsistence payments. Thus we discovered groups of Peterhead building workers commuting to Aberdeen in a shared car and Aberdeen workers travelling to Peterhead. All thereby collected between £23 and £27 tax free a week, an increment that more than offset the cost of running a car.

Housing

I described in Chapter 2 how the planning of housing in Peterhead took place in conditions of such uncertainty that the estimates of the actual numbers needed varied widely. It was also expected that housing resources would be amongst those under the most severe pressure. How, then, did the oil boom affect Peterhead's housing situation?

Three kinds of need were foreseen: (i) the immediate provision of housing for priority workers, in other words, those who were to make the boom possible needed houses first, and quickly; (ii) temporary provision for migrant workers, mainly in the construction phase of both the oil industry itself and in building to meet need; (iii) housing for the increased population. The meeting of need (i) was seen largely as a function of the Scottish Special Housing Association (SSHA); need (ii) of the construction companies themselves (in consultation with the District Council). The local authority, the SSHA and the private market were to cope with (iii), although there was some uncertainty as to the relative share of the public and private sector needed to cope with general needs and the demands of managerial and executive personnel.

(i) *SSHA*

The SSHA embarked upon a scheme of house-building in the Meethill road, totalling 404 houses in 1974. The first houses were due for completion in mid-1975 and all were to be completed by the end of 1978. By mid-1978 the total SSHA stock in Peterhead was 580 dwellings, including 242 built prior to 1975. The highest priority on the waiting list is 1 Oil 'Essential incoming workers to the industries related to the exploration for, or the exploitation of, off-shore oil *or* to the construction industry in the areas affected by off-shore oil developments'. Next is Category 1 'essential incoming workers to manufacturing industries', Category 2 is other essential incoming workers and 3, all other applicants. Applicants are classified continually according to those priorities assigned by the SDA. In May 1978 Peterhead had the following waiting list: the first priority '1 Oil' included 21 names, the second '1' 4 names, '2' 6 names and the lowest category, 3, 112 names. This may be compared with Aberdeen's much more acute 559, 72, 236, 569 at the same date. The Peterhead development had been based on the assumption that the Scanitro ammonia plant would be built. In July 1978 there were nearly 30 empty SSHA houses in Peterhead, 20 of which had been never let. The 7 flats that were vacant had been empty for some time because people offered these small dwellings knew that they would in due course be offered a house. The SSHA was therefore writing to people on the Aberdeen waiting list offering them accommodation in Peterhead. It was not thought that this would produce many tenants.

A 1 in 5 sample of the SSHA allocation made in Peterhead between the end of 1975 and November 1977 (but mainly mid-1976 to mid-1977) yielded the following findings. Forty per cent (12/25) of the allocations were made to the people living in Banff and Buchan, although 4 of these were living in camps. Eight of the local allocations (15 per cent of the total) were made on grounds of overcrowding. This means that a substantial part of the SSHA provision was being used to meet wholly local needs. If we discount people in camps as not being gen·uinely 'local' then 33 per cent of SSHA accommodation went to locals, which would mean relieving the local waiting list by one-third of 404 applications, namely 135 applicants.

The relatively plentiful supply of housing made it easy for incoming oil-related firms to offer housing to their employees. All the companies working at the bases mentioned the lack of problems and ease with which SSHA housing could be obtained. 42 per cent of the sample of SSHA allocations were made to people obviously connected with oil. A further 9 (16 per cent) were in construction and non-oil manufacturing.

Thus the efforts of the SSHA have resulted not only in a relatively easy development for oil-related firms but a contribution also to the housing provision for the local population. The SSHA was not, therefore, meeting the needs of economic growth only but the longer-term housing needs of the town. This windfall was largely the outcome of the failure of development by the petro-chemical industry. The only factor likely to deter locals from taking SSHA houses was the level of rents — a 5-room (6-person) house and garage costing £50.12 a month for rent and rates. This apparently low rent is high compared with the rents paid for older, rent-controlled dwellings in the town, which could be as low as £18 a month. It is appropriate next to turn to the question of the provision for population growth and general needs before dealing with the temporary accommodation question.

(ii) *The private sector*
Private building grew apace in and around Peterhead from about 50 houses a year in the 1960s to 95 in 1977 and 78 in the first six months of 1978. In all 467 new houses became available in the town of Peter-

head in the private market between 1972 and 1978 but the immediate impact of oil was to raise private housing prices. For example, Total Oil Marine entered the private market to provide for some of its workers. It paid an average of £15,100 in July 1974 for houses that had cost an average of £11,780 in May and June and a further £109,000 for seven houses (average price £15,600) which cost £33,725 (average £5,620) in the late 1960s and early 1970s, including paying nearly £18,000 in 1974 for a house which cost £1,000 in 1966. This one company must have been responsible for a considerable measure of the inevitable house-price inflation.

If a trend can be detected in house-sale prices it could be expressed in this way: between March 1972 and March 1974 a house selling for £5,000 at the earlier date would rise to £10,000. In one road where 17 new houses were purchased in 1974 and 1975, 4 of them were resold, the price increases were as follows: £2,525 in 5 months; £2,555 in 6 months; £1,255 in one month; £3,275 in two months. An unpublished survey found that in 1974 the average price of houses sold in Peterhead was £11,170 compared with a Scottish average of £9,775. This finding is not, however, unequivocal proof of higher prices in Peterhead, it may only indicate that houses at the more expensive end of the scale of prices were being purchased. Once again, however, we need to note changes prior to oil; Peterhead was not insulated from national price rises in housing and in the late 1960s and early 1970s more expensive housing was being built in the burgh. Houses costing £10,000 or £11,000 were in themselves new to Peterhead and, although the impact of oil may have been to inflate these prices, the prices themselves already represented new developments 'up market' in Peterhead. In other words, part of the inflation of house prices can be accounted for by more expensive types of houses being built in the town.

(iii) *Council housing*
If the chances of young first-time buyers obtaining a new house in Peterhead were receding, what were their chances of obtaining a council house? In accordance with the policies we outlined in the previous chapter the old Burgh Council built very few houses after 1970: a total of 6 council houses were completed between the beginning of 1971 and local government reorganisation. A further 22 houses were completed after reorganisation in 1975 and 98 in 1976, 118 in 1977 and 207 in the first six months of 1978 (with a further 66 expected to be completed by the end of 1978). During the period 1968–74 92 houses had

been demolished and some 48 closing orders placed on houses. The fact that in its latter years the Burgh Council had pulled down more houses than it built was a matter of pride for the convener of the old Burgh Housing Committee.

From 1976 onwards the supply of council housing expanded rapidly and in 1976–8 the local authority housing output was over half that of the previous decade. The plans for this expansion were made in expectation of petro-chemical developments which did not take place and so the houses served the needs of the local waiting list more exclusively than might have been expected. In August 1978 the Peterhead waiting list was 281, but this included 76 young people not married but accumulating points for waiting, in anticipation of marriage. The actual waiting list therefore was 205.

The waiting list is not the most accurate guide to housing need because potential council house tenants have to recognise their need and entitlement and even then they may not put their names down for a house if they believe there is no chance of obtaining one. In other words demand may be a function of supply. For example, between June and August 1978 the District Council rehoused 20 families and during this period the waiting list lengthened by 75 — the date at which one counts might affect the judgement one would make of the council's performance. The SDD estimated the waiting list to be 126 in 1970, rising to 390 in 1974, although we do not know if the SDD's figures exclude those not deemed to be in real need. By late 1977 the waiting list seemed to be back to its 1970 level with something like 400 families rehoused plus about 110 local families rehoused by the SSHA. The fact that the 1970 waiting list had been housed four times over and remained the same may either indicate the extreme inadequacy of Peterhead's prior provision (not an unreasonable deduction, given its housing policy), or the generation of a demand for housing by the dramatically and visibly increasing supply, or by an increased rate of household formation, resulting from a reduction of emigration amongst the young. The changing characteristics of the waiting list are difficult to demonstrate given the lack of detailed historical records but in 1970 it included no incomers because of the need for residential qualifications. In 1970 about 34 per cent of the waiting list comprised older people awaiting 2-apartment houses* but in August 1978 this was down to 13 per cent. The impression in the housing office was that the waiting list was less

*A two-bedroomed house.

local, younger and had few long-wait members. It was said that the list
of those in 'real need' would be cleared by the end of 1978 when the
current house-building plans came to an end.

TABLE 10

	Local authority	SSHA	Private	Total	Cumulative (%)
1960–5	589	–	202	791	25.2
1966–70	349	200 (in 1969)	295	844	52.1
1971–5	28	42	299	369	63.9
1976	98	102	51	251	71.9
1977	118	130	95	343	82.8
Jan.-June 1978	207	170	78	455	
(est. 1978)	(273)	(190)	(NK)	(c. 541)	100
Total	1,455	664	c. 1,020	c. 3,139	100

The expansion of housing supply from all sources can be seen in
Table 10, from which we note that over a quarter of the new housing
stock built since 1960 was built in 1977 and 1978. New building was
equal to 31 per cent of Peterhead's 1970-1 housing stock. This figure
alone gives an idea of the scale of expansion of the town of Peterhead, but
it has also to be contrasted with the fact that 20-25 per cent of Peterhead's housing stock was below tolerable standards in 1970.*

The housing plans for Peterhead underwent considerable modifi-
cation: *Peterhead '73* suggested that according to the draft structure
plan most of the residential developments would be to the northwest of
the town. A 'green wedge' would then separate housing from land zoned
for industrial use in the Dales farm area. Up to 1976 housing was to be
built in Hill of Grange (120 houses), Copelandhill (160), Middle Grange
(300) and possibly Clerkhill (250), Mile End (130) and Invernettie
(100). In all 1,000 houses were to be built.

In June 1974 Aberdeen County Council purchased the Prison Farm
for £614,000. About 400 houses were proposed to be developed by
Peterhead Burgh Council, the SSHA and the County Council. Towards
the end of 1975 tenders were accepted for the building work and then
the Copelandhill development was cancelled. This latter cancellation

---

*The survey which established these data cannot be found anywhere in local
authority or private records. It has been claimed by some officials that the
figures refer to *town centre* housing only.

arose from the realisation by the then new Banff and Buchan District Council that the developments at Peterhead, Mintlaw and Cruden Bay were likely to be well in excess of needs. The Copelandhill site was serviced by the time of the cancellation but this left the council in a position to make an immediate start on house-building should the need arise. The original plan for over 1,000 houses had therefore been reduced to about 700. The need for such a substantial reduction can only be attributed to the excessive expectations aroused in the County Council by the 'oil boom' in the conditions of uncertainty described in Chapters 2 and 5. As we saw in Chapter 2, in 1973 anything from 2,500 to 3,000 houses were proposed for Peterhead. The diminishing possibilities of work beginning on the Scanitro undertaking probably underlined the extent of this excess.

(iv) *Temporary accommodation*

Aberdeenshire County Council made a type of housing provision that is quite rare for a local authority. They erected 150 chalets at Keyhead, between St Fergus and Crimond. The intention was to provide housing for the families of longer-stay personnel at the St Fergus site. The chalets could either be used to accommodate families for the duration of the job for which they had migrated or until they were able to buy a house or perhaps qualify for SSHA accommodation.

The chalets were well-made wooden prefabricated structures, lined with formica and standing upon properly serviced plots in the estate. There was no enclosed space outside them although some residents erected fences and gates. The site itself seemed pleasant enough on a fine day, with views of open countryside and a wide expanse of sky — it might almost have been a holiday camp. In foul weather, however, the camp was exposed to the full force of wind and rain and the grass around the chalets turned to mud. The site had one small shop and an irregular bus service. Crimond, also with one shop, and the primary school were 2 km to the north and Peterhead 12 km to the south. There was initially no provision made on site for children; this seemed short-sighted given that a migrant population was likely to be young and therefore including a high proportion of children. This was later remedied with a £6,000 landscaped play area.

Originally it had been intended that the local authority should not use the estate for normal needs, but then it was decided to give 30 chalets to local applicants, and when I first surveyed the camp in February 1977 33 per cent of the residents were there because they wanted housing, and nearly half of the residents (46.4 per cent) had previously lived in the Buchan and Gordon districts. At the time of the final surveys in April and June 1978 over half the residents were there simply out of housing need. A number of the local incomers at the first survey were ex-farm workers who had been evicted from tied cottages or were eager to escape from poor agricultural housing. It seemed in early 1977 that the use of the estate for local purposes was inevitable. The housing department had in Keyhead a ready supply of 'temporary' accommodation situated out of sight of the local communities. When pressed by the social work department to accommodate homeless or problem families, the temptation to offer Keyhead chalets would be irresistible. On the first survey I found a number of single-parent families, five of whom were women with children who had been thrown out of their Peterhead home by a husband who wanted to move a new woman in. In mid-1978 such families were very much in evidence still and had been added to by two retired agricultural workers from the Western Isles who had been told they had no hope of housing near home but that they could find somewhere in Peterhead.

I have referred to surveys of Keyhead; the first was carried out simply as part of this research. The second, using our questionnaire, was done by a community worker and the third — again using our simple 19 question schedule — by students of Peterhead Academy under the direction of a sociologist teaching at the Academy.

In all three surveys there was difficulty in establishing which chalets were occupied, which abandoned but formally let, or let and informally changed hands and which were simply empty. Each survey was conducted by visiting and thoroughly revisiting the estate: our contact and interview rate was good, in the last case being 100 per cent as shown in Table 11.

TABLE 11

	Chalets occupied	Contacted (%)		Interviewed (%)	
February 1977	113	95	(84)	86	(76)
April 1978	54	50	(93)	49	(91)
June 1978	36	36	(100)	36	(100)

In February 1977 46 per cent of the inhabitants were locals: this rose to 59 per cent in April 1978 but had fallen to 44 per cent by June because of rehousing by the District Council. The occupancy rate of the estate seems to have fallen to one-third of the February 1977 level and the total population had fallen from about 300, of whom about 120 were children of school age or below (40 per cent of the population) to 131, of whom 63 (48 per cent) were children of school age or below. This is to say that nearly half the population of the estate has always been children. At the time of the first survey 14 per cent of these children lived in families with four or more children but by mid-1978 62 per cent were in such families; 25 per cent of all households had four or more children in them. Larger families, it seems, were remaining in the estate and there were 5 single (female)-parent families at the first and last count and 12 in the April 1978 survey. The estate was, in general, young with 54 per cent of the population being under 30 and 79 per cent under 40 in February 1977. In 1978 49 per cent were under 30, 25 per cent under 40 and 25 per cent over 40.

In early 1977 61 per cent of the families there had come in pursuit of employment, either having a job or seeking one at St Fergus. One-third of the inhabitants were in Keyhead because it offered housing. By June 1978 only 42 per cent were there primarily for reasons of employment and 50 per cent for housing. At the last survey 69 per cent of the households had applied for council housing, 19 per cent intended to remain in Keyhead and 11 per cent had no plans (compared with 4 per cent with no plans in 1977).

The other striking change in the population was its occupational structure. In February 1977 professional and managerial workers made up 24 per cent of the population but by June 1978 this had fallen to 6 per cent. The unemployed had increased from 5 to 16 per cent and two retired households had appeared.

The professional and managerial staff who were 'old hands' in the construction business found Keyhead an improvement upon the small caravan in a muddy field for a high rent which was their usual experience. Large chalets on serviced sites were something of a luxury and, being used to mobile homes, they knew the techniques for maximising comfort. Others, however, moving from traditional housing without any instruction upon chalet-living fared less well. In many chalets we found 2 kilowatt heaters in the living room (usually of the imitation log-fire type). These created a steep temperature gradient across the chalet which with glossy walls created heavy condensation on bedroom walls. 'Damp' was the main complaint. Old hands ventilated their chalets and in some cases installed extractor fans in the kitchen and ensured an even temperature throughout the chalet. They had no dampness problems.

But it was outside where the most complaints originated; mud, litter, broken glass, vandalism. There was nothing for the large number of children to do outside school hours, empty chalets were damaged and fires started. When, as a result of pressure from the Keyhead Community Association, fire-points were installed in the estate roads, they were wrecked by children. A pack of dogs also developed — many households had dogs for protection and some of these dogs were abandoned by departing residents. These dogs were a hazard to the remaining population as well as to sheep on the surrounding farms and contributed to the degradation of the environment by ripping open plastic rubbish bags and rooting for food.

A series of attempts were made to run a community association to provide amentities: playgroups, bingo, homecraft classes, youth clubs, etc. The association ran sporadically and suffered from losing personnel with population changes. The chalet obtained from the District Council by the association was attacked by vandals. The most tangible results of the KCA's activities were the fire-points installed after one chalet had burnt to the ground before the fire-brigade arrived, and the rerouting of the bus into the estate so that the children did not have to cross a road busy with construction traffic in order to go to school in Crimond. The shop and the telephone box were also obtained by KCA pressure.

Perhaps the most obvious problem for Keyhead was the growing stigmatisation of those who lived there. In 1978 respondents said they could not get jobs because of their address. The housing of 'problem' families in the estate started the stigmatisation. This was added to by knowledge of and rumours about the irregular family lives of a section of the construction workers — both incoming and local women living with workers and changing partners when their original man moved on. These 'goings on' were amplified by rumour and gossip and by the appearance of Keyhead addresses in the court lists alongside charges for stealing, drunkenness and violence. The stigmatisation reached its most acute in mid-1978 when a caravan camp containing a number of prostitutes was disbanded and some of the prostitutes went to Keyhead. The women residents felt that they were all classified as loose women, and the single women felt especially vulnerable. These respondents, when asked questions by the Academy students, mentioned naked women fighting and shouting in the open at 2 a.m. (one respondent alleging that they were 'covered in tomato ketchup').

By the time of our last visit to the estate most of the chalets had boarded-up windows and up to 50 of them were due to be moved away; the whole site seemed like an abandoned camp. Residents reported themselves unwilling to go out in case their homes were robbed, the lack of public transport and any amenities made them feel isolated and the lack of transport also reduced their chances of finding a job. It had been widely recognised throughout the life of the camp that it was a 'dumping ground' for problems, and this further isolated the 'respectable' residents. None the less, one or two families could always be found who said that this was a lovely place to live, somewhere in the country, better than they had before. In the last survey one respondent volunteered the opinion that Keyhead was very much better than Dundee council property.

The possibility of the estate's becoming a dumping ground had been recognised from the beginning, because Keyhead offered a supply of ready-use housing which could be used without provoking resentment amongst people on the waiting list. But it became more than this; it was holding places that could be used for sorting out those who were 'suitable for a proper council house' (local official). In effect it became a 'dumping ground for undesirables' according to one housing official. This view was recognised by the inhabitants.

This view, namely that the estate was a 'dump' rather than a 'transit camp' can be seen from the fact that whereas in February 1977 only 1.2 per cent had been in the estate at least 33 months by June 1978 14 per cent had. Although the actual number of long-stay households (over 15 months) had been reduced from 50 to 16, this latter group represented a hard core of long-stayers who had not been rehoused and who were the parents of 60 per cent of the remaining children.

In practical terms, therefore, Keyhead may have been a success in providing housing for key workers when none other was available. In social policy terms it was a disaster because it created a demand by being a stock of housing in the vicinity of a rural population in housing need. It lacked the amenities to be used as regular housing and, by virtue of its use as a temporary location for persons creating problems for the housing department, it was stigmatised. The high population turnover, isolation, lack of amenities and bleak location sapped what little morale and 'community spirit' there may have been, and which would have been the basis for a degree of social control which might have reversed the demoralisation. Banff and Buchan District, a mainly rural authority, managed to create all the features of inner-city deprivation in the countryside.

(v) *Camps*

The Keyhead Estate was the least successful form of local authority housing provision encountered in the course of this research. The work camps provided for itinerant workers had been of prime interest in beginning the research project. This part of the study was relatively neglected, however, when it appeared to be much less important than other considerations.

Three labour camps were built at St Fergus, one housing 170 men and two housing 400 each. (This compares with the Boddam camp, which housed 680 men.) The contractors preferred separate camps for their men. So strong was this preference that they would rather build their own new camp than share an existing one. The reasons given for this varied; the junior managers tended to say that having separate camps gave the company control; they knew how much accommodation there was and could control its allocation. More senior managers said that there were 'industrial relations' reasons. This was elaborated upon by Mr Lines of Shell at the NGL enquiry when he said that men in one camp could compare notes and that 'differential conditions between one site and another can cause industrial unrest' (*Enquiry*, p. 1097). Our impression was that civil engineering workers would tolerate worse conditions than mechanical engineers.

It was expected that 90 per cent of the construction labour force would be temporary and therefore housed in camps. The most important unknown was the extent to which local labour would be recruited and the need for camp accommodation reduced thereby. One of the 400-men camps never had more than 250 men in it, and neither was ever full during the Total/BGC contract.

The formal planning requirements for labour camps were laid down by Banff and Buchan District Council in a document entitled *Accommodation of Migrant Workers*, which was published after the Boddam and St Fergus camps were built. This apparently extraordinary omission was because no camps were included in the original proposal. The planning requirements were aimed at the Scanitro and Shell/Esso developments in the light of experience at St Fergus. Paragraph 2 suggests that camps should be near urban centres so that workers could use leisure and social facilities and it also suggests that camp recreational facilities might be used by locals to facilitate local-migrant integration. Barbed-wire-enclosed compounds give a 'concentration camp' appearance and should be avoided.

The Crimond camp site is a kilometre from the Fraserburgh–Peterhead road and is reached by way of a muddy, potholed track. The site is exposed and the walk to the road is daunting in the rain. Crimond, 2 km away, has no public facilities. There is a bar at Keyhead (2 km) and St Fergus (6½ km). Each camp is separated from the others by a high wire fence. In the early days of the camps no buses were provided to Peterhead, the idea being to prevent the sudden descent of large gangs of men upon one pub and to encourage the orderly drift of workers to town by public transport. This created extra business for the town's taxis.

Civil engineering workers do not, on the whole, stay long on a site. We have already noted that 50 per cent stayed less than three months. The camps therefore did not house men enduring prolonged periods of isolation in utterly remote areas. Whilst 'integration' with the locals may be a concept devoid of any concrete meaning, the camp-dwellers did develop a taste for the local recreations. Buses were run, on request, to Fraserburgh, Rosehearty and other locations for dances, and construction workers were to be found in many Peterhead pubs towards the end of the week.

The small camp was run by a local manageress after two camp bosses had 'become too friendly with the men' within weeks. As a result of this local management food orders were placed with local dealers and the manageress was able to select her local female employees with care. The camp workforce was not unionised, and employed 13 women and 3 men. As in all camps men could be expelled for having women on site, at one time both prostitutes and teenagers from Fraserburgh and Peterhead had to be chased away from the perimeter fence by the manageress — but no local girls were there because they knew their families would be informed. The camp also forbade spirits and allowed only beer in the bar. The other two camps were much larger, one run by commercial caterers and the other by a contractor who resumed control of his camp after commercial caterers refused to re-negotiate a contract which had initially assumed a very much higher population in the camps.

At some stage in the career of the research project all the researchers had occasion to eat in camps and we found the food to be excellent and served in quantities that relatively sedentary academics found quite unmanageable. The main problems in the camps consisted of labour problems for the catering contractors and drinking problems amongst the construction workers. A 400-man camp employs a staff of about 40. There is a high turnover of staff with a lot of 'cowboy' cooks with no qualifications trying their luck for high wages, or hotel chefs with no flair for industrial catering. Drunkenness is also common amongst catering staff and the Health and Safety at Work Act forces management to shed such workers very quickly. Cleaners and chalet-maids were easily recruited locally; their hours could be adjusted to suit domestic circumstances — all the camps said that local women were 'good workers'. The manageress of the small camp also provided a free crêche/play group under professional supervision in the summer.

The camp managers tried to separate their activities from those of the construction site. But this was not always possible: one manager reported evicting a man for fighting in the camp and then coming under pressure from the man's employer because he was a key worker. Another manager was ordered to close his camp when the men struck on site, but the order was cancelled two hours later when the employers appreciated how much long-term damage could be done by such a provocative action. The more experienced professional camp managers seem more able to resist their clients' demands in this respect.

Wages and conditions cause two kinds of problems. Firstly, the difference in terms and conditions of locally recruited and migrant catering staff and secondly, the differences between civil and mechanical workers. The first difficulty can be resolved — even when to do so is contrary to wages policy — by paying local people a living-out allowance and the non-locals a living-in allowance thus buying off both sides (a practice with which we were thoroughly familiar on the construction sites). One camp operator eliminated wage and demarcation problems by employing all classes of worker as a 'general hand'.

The mechanical engineers had longer weekends than the civil engineers and higher subsistence. But the camp charges were all the same, so the men on higher subsistence pocketed the difference (tax free) and this was a source of some resentment. Such resentments no doubt surfaced under the influence of the major problem, alcohol. The isolation of the men and their relatively high wages made drink a problem within the camps. Drink advertisements stress the manliness of drinking, and drinking as a male activity. Television advertisements regularly use the image of physically hard working men coming together for a drink after work — oil rigs, service vessels and the Boddam power station all feature in the presentations. Drinking is a major problem in Scotland, with five times the rate of alcoholism as in England. All social gatherings (except those of a religious nature) tend to be accompanied by the consumption of considerable quantities of intoxicating beverages. This is as true for students spending an evening with friends as for construction workers in the camp canteen. The difference is that the students have less money and more   alternative activities. Heavy drinking was reported as a major problem by all camp managers and some of the employers. Estimates varied but managers reported between 10 and 25 per cent of the camp dwellers *not* being heavy drinkers and even the very steady older men occasionally 'going off the rails'. These estimates may well be exaggerated, but they confirm one another. Two main results of drinking are bed-wetting in the camps and absenteeism on the site. Men habitually absent were normally sacked, and construction companies reported very little drunkenness on site, some of that being amongst locally recruited men.

Visiting managers from English firms found conditions in the camps quite amazing. One trouble-shooter for an English firm said that he had been in the construction industry for 20 years but was 'amazed at the Scots living in the camps here. They are like animals at work and leisure, with squalid problems with drink and bed-wetting.' He reported a dispute in his firm when the men complained that the Friday Glasgow bus was not leaving punctually at 3.30 p.m. and thus not arriving in time for the last trains to Glasgow. He personally got the bus away on time the following Friday and then at 5 p.m. he found it in Peterhead awash with beer and vomit, loaded with crates of beer and unlikely to get to Glasgow before 11 p.m.

This problem is, however, internal to the camps and does not confirm fear expressed locally about drunken navvies rampaging in the town. In more general terms, the camps had little effect on the local community. In Peterhead they seemed quite remote and the St Fergus operation was only discussed when there were industrial disputes there. Of course construction workers could be found in the pubs, but they behaved like any other citizen out for the evening. In Crimond the impact was initially nil — perhaps because Crimond lacked a pub. The district was not new to migratory camp-dwellers, having had the Fleet Air Arm and the WRNS in the war and squatters after. Some construction workers played football against a Crimond team and some helped with the Gala. The impact of the camps upon Peterhead may have been important in terms of providing work for women and putting cash into the tills of bars and cafés but in social and political terms their impact was very slight.

The camps we have described so far were purpose-build, static installations. There were other camps of a less permanent kind. Experience with caravans in the area had been confined mainly to holiday traffic, for which regulations relating to space, water and toilet facilities had been developed. Caravans and other temporary accommodation were allowed on a construction site – permission to have them is presumed when permission is given to construct. A number of small house-building sites had caravans and chalets, as we have seen, but much larger sites appeared at Longside aerodrome during pipe-laying operations. In Table 9 above we saw that inshore pipe-laying employed about 400 people. This operation moved across country in 40-mile steps, setting up pipe-stores and work camps along its route. Longside was a pipe-store and the contractors assumed therefore that they had permission to park caravans – even though they were *storing* rather than *constructing* on site.

The Longside caravan site became notorious as the residence of a group of incoming prostitutes. At one time it contained 66 caravans ranged down two sides of a portion of potholed runway. The caravans were of all shapes and sizes, ranging from luxury mobile homes to one-man tourers. They were surrounded by gas bottles and some had tarpaulins, lean-to's and wires to secure them – a testimony to the exposure of the site, which afforded a 360-degree view of the bleak Buchan countryside. It was the worst kind of uncontrolled development encountered in the whole of our study. But its effects seem to have been minimal. The conditions for the inhabitants were obviously very poor and considerable health hazards must have been posed by the lack of services.

The only other kind of caravan sites were the small private ones situated behind a garage, in a field, beside old buildings, etc. These were strictly unauthorised and became the object of legal action by the District Council. The considerable profits to be made – up to £30 a week for the hire of a small caravan – probably made it worth the offender's while. They were not studied in the course of the research, and had been removed by the time this research was completed.

'Social problems'

We have mentioned drunkenness and prostitution. Some Peterheadians also expected that migrant workers would include criminal elements who would add to the demoralisation of the area and 'culture shock'. Did Peterhead suffer a wave of lawlessness and immorality? The answer

is fairly simple; there were increases in specific areas of delinquent behaviour but not attributable to oil, there was a slight increase in prostitution in a seaport that already had a well-established trade, and drunken
disorder declined. Between 1973 and 1974 cases of theft known to the
police increased fourfold from 48 to 267, and housebreaking almost
doubled as compared with a 15 per cent and 17 per cent national increase in these crimes respectively. In the opinion of the police these
increases were the result of the activities of a small group of local youths
known to them. Breach of the peace also increased and some of this was
attributable to a greater willingness to call in the police to cases of
domestic violence. The 'rolling' of men, enticed by women from bars to
their homes in certain districts of the town where their 'husbands' set
upon the men and relieved them of their wallets, featured in the Sheriff's court list regularly. Such activity is of great antiquity and began
when the first seafarer sought a bar and a girl upon coming ashore.

Prostitution seems to have increased but in the nature of the trade it
is not possible to conduct statistical analysis, especially as there is
likely to be a fringe of women who may bestow favours in return for a
'good time' without adopting professional status. Full-time prostitutes
from Glasgow did move into Longside and the town, but there was
little open soliciting. Certain bars were alleged to be pick-up places but
our observations did not confirm this. The 'all-night' charge for a prosti-
tute's services was £40–£50. Stories of schoolgirl prostitution were also
common, but we only found one such schoolgirl and it seems likely
that she was the case which everyone knew and which gave rise to all
the gossip. Also in a fairly straitlaced community, where children have
been kept under close control by parents and lack of money, the new-
found affluence of some of the youth and the associated life-style could
easily lead to allegations of 'loose' behaviour by girls emulating the
established behaviour of their age-mates in the more affluent regions.
Doctors responsible for treating venereal disease reported a ninefold
increase in cases between 1970 and 1976 and attributed this not to
prostitutes – who are careful – but to 'the many willing amateurs' who
spread infection. Not all of the increase, however, is likely to have been
local but the result of an increase in harbour traffic bringing more men
to Peterhead for treatment. Venereal disease in the camps at St Fergus
was mainly contracted in Fraserburgh. One construction firm found 8
of its 40 men had VD, but this could have been the result of only one
common contact. We were not convinced that Peterhead was becoming
a northern Babylon.

The Spanish and Italian workers from the lay-barges which used Peterhead in 1973 were held up as models of good behaviour — almost as if they were expected to be especially delinquent. One reason for good behaviour was that the employers sacked anyone who caused any kind of trouble in port. One Italian allowed himself to be driven at knife-point off a quay and into the harbour rather than put up a fight and cause 'trouble'. Only intervention by 'respectable' locals after the event prevented his being sent home by his employer.

Some Spaniards and Italians married local girls and both the Catholic priest and a Presbyterian minister reported marrying couples either with an interpreter or in a language that one partner did not understand.

The police reported an improvement in street behaviour with the reform of the licensing laws in 1977; people seemed to trickle home from 10.30 onwards instead of being thrown out of the bars on to the streets at 10.15. This appears to have been a nationwide trend.

What we have described so far sounds quite the opposite of underdevelopment or exploitation. Employment increased; there were higher wages, and an improved housing stock. On the negative side we only have higher house prices, some unsightly camps and a few additional prostitutes. Is this the total impact on Peterhead? Two further questions need to be explored; what happens to the poor and underprivileged in conditions of increased incomes and, secondly, to what extent do local people use increased wealth to develop local enterprises which will survive the construction boom?

The old would be at a disadvantage as affluence increased, if prices increased proportionately. In fact if prices rose faster than average the elderly would be falling behind the average old person; if at the same time standards of living were rising then the elderly's relative position would be worsening in the local community. We found this to be the case. Peterhead has developed a wide array of institutions for helping its older citizens; charitable trusts, organisations to deliver Christmas parcels, meals on wheels. The fishermen occasionally auctioned barrels of fish to raise money and prisoners in Peterhead prison helped cook for the elderly. The Masons and Rotarians raised funds for the old. Yet we found what the fisheries officer had reported in the pre-war depression, a dogged resistance to 'charity'. We both interviewed and were told of people who refused all help offered by the community, who refused even concessionary bus and rail fares, and many who would not draw Supplementary Benefit. Keeping up appearances is very important.

We interviewed one old couple who had to budget for months in order to buy Christmas presents and yet refused the Christmas dinner provided by the Aged and Infirm Committee and the Rotary Christmas parcels as 'charity'. This outlook was used to justify why the old Burgh Council had not adopted a rent rebate scheme.

What we found amongst the old in Peterhead is not uncommon in Britain. It is a major social policy problem to persuade those entitled and in need to take up the available benefits. Costs rose in Peterhead in a way that put old people at a considerable disadvantage, and one pensioner who worked part-time in a shop reported how old people were buying smaller quantities of even quite cheap foodstuffs while 'oil men' and their wives bought only the best and hardly ever asked the price. Younger people could afford shopping trips to cut-price stores and buy in more economic quantities: the elderly immobilised by high transport costs had to pay town-centre prices when they knew these to be higher than elsewhere or in Aberdeen. The prices charged by these shops were widely believed to be exorbitant, as we will see when we describe the events surrounding the Tesco supermarket below. Old people said they noticed how goods which they expected to cost 1 or 2 pence more as a result of inflation cost 10p or more extra in Peterhead. The local offices of the Department of Health and Social Security reported an increase in exceptional needs payments of 20 per cent for the year ending April 1976 and cash grants going up 100 per cent in the following year. About two-thirds of these payments went to the old. Thus the needs of the old as measured by the DHSS had increased sharply and this must have created very considerable strain for those who did not claim benefits.

Another group we investigated was single-parent families. Those located in Keyhead were faced with fares of about £1 if they did not wish to shop in the estate's only shop. The buses from the outlying districts also seem to be timetabled to suit workmen's hours, thus forcing housewives to sit in town drinking tea when they wanted to be home. Transport was a vexed problem; the single-parent family group virtually collapsed because of the lack of transport. Some of them contrasted this with the number of new cars in the town. The younger women especially noticed how prices were cheaper in Aberdeen, but the cost of travel was over £1.50. The transport costs were the reason given by shopkeepers for charging higher prices than in Aberdeen, although some simply denied that they made higher charges.

The higher cost of housing also prevented young people entering the local housing market and we found cases of couples who had not been able to keep up mortgage repayments and had sold their house and put their names down for SSHA or council housing. The poorer housed people in Peterhead were benefiting from the rents freeze and this deterred some from taking new council housing, especially if their current accommodation was within walking distance of the town centre. We also found the social work department concerned about families at Keyhead and Mintlaw because it was felt that those likely to be in the greatest need of their services would not have access to them because of the cost and irregularity of transport. One housing factor indicates what might happen in the future: many people housed by the local authority, especially in Mintlaw, were unemployed. If male unemployment rose after the major construction works were finished and the town established as a service centre, there would be many couples in modern houses who had enjoyed a relatively high standard of living but were no longer in work. If they moved away to seek work they would lose excellent housing; if they remained behind they would have good housing but little else.

So far in this chapter we have discussed the changes in Peterhead from the sociologist's point of view, examining alteration in the economic and social structure which might demonstrate one theory or another of the impact of oil. But what concerned the people of Peterhead in terms of the meetings they attended and the letters they wrote to the *Buchan Observer*? Once the Scanitro protest was over it is very clear that Peterheadians were not mainly concerned with the social impact of oil and they were also a little bored with answering the questions of those who thought they ought to be.

Three issues seemed more prominent than others. The first concerned the physical condition of the town centre. Years of inability to cope with gap sites and a major fire in 1977 had made the centre of the town very unsightly. In addition, heavy traffic through the town, including lost lorries en route to construction sites or bases, had begun to break up the pavements. The roads were also more dangerous to cross, especially for children and old people. Furthermore litter began to accumulate on the streets and refuse collection services were said to have deteriorated (especially by ex-burgh councillors). This was attributed to the Banff and Buchan District Council taking away Peterhead's new refuse lorries and mechanical road sweeper at local government reorganisation.

Secondly, and closely connected, was the question of the Drummer's Corner. This was a prime town-centre site, used as a car park and covered with mud and potholes, scheduled for development. During the last years of the Burgh Council a number of schemes were discussed and were found to be inadequate. In 1972 Maxwell Properties proposed a scheme which was voted out in 1973; discussion of a plan by Telegraph Properties (Scotland) Ltd began in 1973; rumours of other schemes ran until 1977 when BBDC approved a scheme by Netherscot, a Dutch-Scottish consortium, for shopping developments and car parking. This decision was then overtaken in public debate by the decision of the Planning Committee not to grant permission for Tesco to build a store in the 'Buckwell House' field (see p. 18). This decision was carried by one vote and was followed by a storm of protest in which two house-wives raised a petition of over 2000 names in a week. The petition received support from the Clerkhill and Meethill areas especially and to some extent can be seen as representing the 'new' Peterhead. The housewives wanted a superstore with cheaper food and cheap petrol, so as to break the monopoly of the town shopkeepers who charged high prices. In simple planning terms the store spelt disaster because it was planned not as a shop for the town of Peterhead but for the whole Buchan area. Its location would service the area well but leave the core of old Peterhead virtually isolated on a peninsula. It would not only reduce trade but possibly attract labour away from the town shops and prejudice the Netherscot development. The *Buchan Observer* opposed Tesco on the grounds that it was another example of big business over-shadowing the small community of honest Peterheadians. A further irony was that a local supermarket owner in Clerkhill had wished to expand for some time but had been refused permission on planning grounds.

The final issue concerned the Academy, the appointment of the Rector, the style of the buildings and the use to which the community parts of the community school were to be put. These then were the issues that exercised Peterheadians in 1976 and 1977. Their concerns were not altogether those of the sociologist.

The entrepreneurs in Peterhead

A researcher returning from a new oil field in Africa reported that in the local population only taxi-drivers and landladies had benefited. These were certainly two activities that flourished in Peterhead. There was a proliferation of taxis in Peterhead, some of them unlicensed and probably inadequately insured. (Indeed, licensing had been effectively abolished as a result of a drafting oversight in the Act reforming local government.) Airmen from RAF Buchan were said to be running taxi-cabs when off duty, although we never met any such drivers, having no occasion to use taxis. Board and lodgings averaged £30 a week in 1977 with £50 the top price and £15–£20 the cheapest in a council house. About 750 beds were available in lodgings in Peterhead.

Becoming a landlady or taxi-driver does not require much capital outlay and little risk is involved if either is a part-time activity pursued in addition to paid employment. More risk is involved in setting up a new business or adapting an existing one to the changing circumstances. The shopkeepers, for example, seemed fairly passive and many of the shops remained with cramped and ill-designed interiors and dull window displays. But some did expand, improve their interiors and displays, improve their range of products and try to attract new customers. The distinction this difference represented is that between subsistence shop-keepers, who used their shops just to provide a living for their family, and businessmen who owned shops. The latter were more likely to take entrepreneurial initiatives because they were interested in expansion and higher profits. Some of the smaller shopkeepers made unwise attempts to gain from local oil-based developments; one, for example, sought a bank loan to enable him to tender for a contract (with penalty clauses) worth three times his annual turnover. The bank manager pre-vented this venture. A number of small businessmen said they felt they were moving into a new world of business when the expertise of bank-ers, lawyers and accountants was more important than native wit and industry. The reluctance of many to become involved with the oil industry was based either on the belief that oil was a 'fly-by-night' activity which would go as quickly as it came or upon the knowledge that incoming companies made stringent demands upon local firms but paid their bills slowly. This is a contrast in business structure; small local enterprises tend to live from hand to mouth using cash payments from one job to buy the materials for the next, whereas big national and international firms handle their accounts centrally and pay every six months. From either point of view it would be unwise to become committed to oil.

We found four examples of local businesses which had responded actively to the presence of oil-related activity and these cover a range of enterprises. The crucial question about all of them is whether they could, in the language of traditional economics, achieve self-sustained growth that would enable them to survive the decline of the construction phase of oil-related activity.

We saw in the first section of this chapter that the most rapidly growing employment sector was services, and that there was an expansion of work in post and telecommunications. There was also a demand for secretarial and clerical workers within the bases, this being primarily a demand for female labour. A young couple — returned migrants to Peterhead — had set up a private employment agency to meet this demand. Between 50 and 60 companies were registered with the firm in mid-1977 and these companies were in the oil industry. Strictly local demand was confined to the need for holiday staff. The work of the agency shed light upon the local labour market and the *modus operandi* of the oil companies.

The agency dealt with local people, mainly female school-leavers and women who had done special courses in typing at the Buchan Technical College (courses were specially set up to meet the demand for clerical labour). In addition a number of girls working in local banks sought to register with the agency because they thought they could earn high wages. It was explained to them that most such jobs were temporary and usually began immediately, so girls would not be able to give the required notice at the bank. Local informants contrasted 'steady' work at the bank with what one social worker described as ' "Girl Friday" jobs, running around barefoot amongst the rubber plants'. There does seem to have been a lot of glamour associated with oil-related work in the early days, with office parties and prospects of travel, but this soon settled into a more regular routine as the offshore developments took a more stable and predictable form. One problem encountered with local girls was their broad Buchan accent which is extremely difficult for foreigners to understand. It was felt that girls ought to adopt a more internationally understood version of English. Incoming women registering with the agency were found to be more skilled than local women, 60-70 per cent being able to type and some being able to operate computers. Such skills had never been in demand in Peterhead in the past and so few had acquired them.

The work offered to agency girls was temporary and enabled the employers to operate a probation period without contravening the Employment Protection Act. The girls remained employees of the agency until such times as they were offered permanent jobs. In effect the oil-related companies had subcontracted the hiring and firing of secretarial staff to a specialist firm and this saved them from having to cope with fluctuating labour demand themselves. This subcontracting activity in itself might have a long future, but at a fairly modest level as the construction phase passes. It provided a service but no basis for economic growth. The founders of the firm took an opportunity that development offered and created a successful enterprise. They opened an office in Aberdeen in 1978 where they entered a much more competitive world of employment agency activity. There were about 30 such agencies in Aberdeen in mid-1978.

Retailing was also expanding and one food store had taken advantage of the economic boom in Peterhead to expand business. The manager was a graduate and an incomer who had married into Peterhead. The store was sited in Clerkhill in the centre of the 1960s housing development and close to the Meethill housing sites, thus it had an assured local trade on its doorstep. Like any other supermarket it offered reductions and loss leaders in staple foods (bread, sugar, cereals, etc.) to bring customers in. The manager adopted a policy of supplying whatever the customer wanted. He contacted the first English-speaking member of each incoming group of foreigners to find out what they needed. Thus by mid-1977 his store had a vast array of pastas, patés, exotic seafood, hams, sausages, cheeses and biscuits from virtually every country in Europe. On our first visit we found quails' eggs available on the shelf. This policy brought new customers in and as a result the manager had been able to expand other parts of his business; his malt whisky stock had risen from 4 to 40 varieties and his wines from 6 to 250 and people were coming from as far as Aberdeen to shop in the store. Local people were gradually trying some of the more unusual foreign foods partly because they were on the shelves and partly because they recognised food from their foreign holidays.

The store employed up to 40, including 7 young people who worked evenings and weekends. The store was ripe for expansion but could not get planning permission to do so because of inadequate parking space. The manager was meanwhile seeking ways of diversifying into furniture and furnishings, using a town centre shop. The profit on these goods is very much higher than on foodstuffs, which can only be profitable with very high turnover. It seemed likely that the Tesco store would take away some trade from the Clerkhill store, although perhaps not local convenience-shoppers. It could not survive solely on supplying specialities to passing Italians and sociologists. Thus this second example shows the exercise of considerable entrepreneural skills in the context of the opportunities offered by Peterhead, but the future of the enterprise was uncertain. The best it offered was the chance to expand retailing to a new plateau.

Our third example is of servicing rather than services. This is a firm of engineering contractors specialising in welding. Thirty-five men were employed 30 of whom were time-served men and 5 apprentices. Only one man had left the firm since 1970. The boss, a time-served man himself, worked alongside the men. When the firm was started in 1969 about 60 per cent of its work was on boats but by mid-1977 it was only 30 per cent. A further 30 per cent of the work was subcontracting work for the major contractors at Boddam and St Fergus, and another 30 per cent on jetty and oil installations in and around the harbour.

The firm was working at full stretch and the men were able to work 70–80 hours a week which gave wages of about £147 a week with £100 for apprentices. The main difficulty was an administrative one; the boss was a practical man and unaccustomed to office work and yet as his business grew so did the office work. The firm's 'business' was held in his secretary's head and the boss was utterly dependent upon her and had installed a small computer to help with her work. But he had reached the point where he needed an accountant and more office staff to keep the invoices, orders, wages and tax affairs moving. He had no space for such an office in his already overcrowded yard and workshop.

About two-thirds of the firm's work was oil-related. It was therefore very *dependent* upon oil and, although it might undertake contract work for many years to come, it would be doing so on a reduced scale and in competition with the BOC base development. With a continuing boom in fishing and the greater use of the harbours by fishing vessels as a result of the enlargement of the market there would be an increase in engineering and fabrication work for boats. The future of the firm needed careful planning and it is doubtful whether in the day-to-day pressures of work anyone had time to plan.

Our final example is of Peterhead's local millionaire. As he was interviewed by virtually every visitor to the town and featured in numerous newspaper articles, it seems unnecessary to conceal the identity of Mr Ferrari, who, by the time we met him, was an old hand at being interviewed.

Mr Ferrari had been in the fish and chip business and owned a betting shop. He sold the bookies' shop to raise cash for a hotel. He claimed that in 1971 he was still frying chips himself, although he had inherited some shop premises, owned a restaurant, two hotels and a bar. The hotels were vital because, using them as a base, Ferrari was able to provide a wide range of services for visiting businessmen in the oil industry, such as changing money, buying and hiring cars for them. He was the man on the spot, who could be contacted in his hotels outside working hours and who had the right local contacts to get things done. Also in the bar he began to listen to oil men talking and to discuss their needs with them. He reckoned that men relaxing over their drinks in the evening had brought him £100,000 worth of business in 1977. There had been a few failures, for example, his attempt to set up a company to build submersibles.

His most successful venture was the Peterhead Engineering Company. This employed 110 men in its engineering works of whom 40 were welders who were able to meet the most stringent demands of American offshore companies. There were also 20 apprentices, 10 staff, electricians, platers, blacksmiths, a joiner, drivers and cleaners. Ferrari was renowned for his high wages and being a demanding employer who expected his men to work at any time of the day or night. He was also a bluff character who would turn up with a crate of beer for men working in the middle of the night. His engineering firm was not unionised but AUEW had an informal consultancy status. Ferrari reckoned by being a good payer he had raised wages all around the harbours.

Ferrari's success was based on his considerable stock of personal energy and initiative. He had properties with which he could raise capital (and more capital as property values rose). He was well placed to gather business intelligence. He was not, originally, one of the Peterhead elite, and thus quite willing to engage in wheeling and dealing and becoming a 'fixer' for incomers. He was very critical of traditional finance houses which he felt lacked imagination and enterprise but with the growth of his activities he, like the boss in the previous example, began to realise that he was entering a world where accountants and bankers were kings and the practical man at a disadvantage. At our last meeting he said how much he would like to know what the oil industry would need next, so that he could make it for them now. The Engineering Company was 90 per cent concerned with oil, onshore and offshore. It seemed to have every chance of survival into the repair, planned-maintenance and medium-scale fabrication activities that would be integral to the production phase offshore. It also seemed the sort of company that might be taken over by a national or multi-national corporation who wished to enter the offshore servicing sector. But in 1977 Ferrari was already exploring the possibility of expansion into Latin America.

It is important to note that, when small firms like the engineering contractor or Peterhead Engineering grow, the 'business' of the firm outgrows the skills of the essentially practical men who run the firms. Thus they have to turn to bankers, lawyers and accountants to advise them; they have to enlarge the office and employ administrators. The firm then undergoes a qualitative change and the entrepreneur becomes much less independent, finding himself subject to new sorts of commercial and administrative constraints. Bankers, lawyers and accountants take none of the risks but benefit from the extension of their interests. It is the banking, legal and accounting services which have experienced the most extensive growth — in employment terms at least — in Peterhead since 1972.

So far we have dwelt upon businessmen as entrepreneurs. But one fear felt by local employers was that the arrival of big national firms in the area would bring big national trade unions and 'trade union attitudes'. We have already commented upon the lack of development of a labour movement in the area and the consequences of this for the population.

Trade unions

We saw in Chapter 5 that the Peterhead Business Association and the Harbour Trustees feared unionisation of the harbours. This was part of a more general fear felt by small employers and businessmen in Peterhead, namely that in addition to raising wage levels, incoming workers would bring 'trade union attitudes' into an otherwise harmonious Peterhead. This 'harmony' we suggested in Chapter 5 rested upon the domination of the burgh by small property owners and an almost complete lack of working-class organisation and representation. What was feared, then, was the organisation of labour in a town dominated by property.

Those who expressed this fear either took no account of the fact that labour had been organising effectively since the late 1960s or did take account of it and felt threatened. During the 1960s the workers were organised in the engineering works by the AUEW (and its predecessors), by USDAW at Crosse & Blackwell and by UCATT amongst the construction workers. The real threat came from the AUEW because it was drawing on a long craft-union tradition to organise workers in the two key non-fish industries in the town, and industries which set the pace for local wage levels.

The AUEW has had negotiating rights for all workers (including 4 electricians and a builder) at General Motors since 1965. Both the union's founder members were incomers and both ex-miners, one from Lancashire and the other from Fife. They found considerable difficulty in organising a union. The share-fishing and 'free enterprise' tradition made people resistant to collective action and the size of the community made them reluctant to adopt combative and partisan positions publicly. As one local AUEW official put it, 'everyone wants to be nice to everyone'. He then went on to describe how in disputes members unwilling to stand up in public told him privately that they supported the union – but perhaps they were just being 'nice' to him. One special problem encountered at workplaces in Peterhead was the presence of the Brethren who resolutely refused to join trade unions. There was then only one left in General Motors and he paid his union dues to charity. The local union branch enjoyed considerable autonomy and staged walkouts over such matters as representation in discussion with the Pay Board and the lack of full-time officials. They also organised an overtime ban in pursuit of parity with the Lanarkshire branch of General Motors. One of the prime objectives of the union leadership at General Motors in Peterhead was to achieve unity with other General Motors workers in Scotland, so that they could all negotiate from strength and achieve parity of conditions.

The union at Clarksons Tools had not had a smooth history. An early union had been broken up prior to 1962 and the manager said in 1966 or 1967 that there would 'never be a union here as long as I am here'. In 1973 he left and went to Fine Tubes in Plymouth where he seems to have been a key figure in promoting the longest running postwar strike (*The Times*, 14 June 1973). The manager was not only following his own predilections in being anti-union but prosecuting the policy of the American parent company.

The union had some success: in 1962 a 15-minute stoppage gained 6d. an hour and recruiting expanded, but the District Secretary did not press for recognition and membership dwindled to about a dozen in 1970. The company remained anti-union until Clarksons took over and 'industrial relations' improved slowly. The union said that consultants reported to the management that industrial relations problems were caused by the workers and this may have coloured the new management's attitude. In 1972 there was a two-month wages strike and at one point management recalled the workforce to negotiate wages, it seemed: in fact they had hoped that by bringing the workers back to the factory they would start them working. The workforce was 'out' again by lunchtime. The situation was described as 'relaxed' in 1977 because there was an industrial relations officer who took the union seriously.

The same could be said of USDAW which in 1970 found itself at the centre of a closed-shop dispute at Crosse & Blackwell, in the course of which the company threatened to close the works. A member of the Brethren drove his daughter through the picket-line at high speed and finished up in court and feeling generally ran very high. The full-time official reported Crosse & Blackwell as 'difficult' and said that in resisting the union the company presented itself as the defender of the rights of minorities (the Brethren) against the union. The 1970 dispute was the culmination of grievances; in 1968 there had been a recognition strike which resulted in 100 per cent unionisation (except for Brethren and RAF wives) and a pay rise. The engineers had also struck and the night cleaners, both of whose wages had improved as a result. One difficulty was that management only told union officials what it chose. But the main difficulties encountered by shop stewards and regional officials were that the company tried to use the General Industrial Council pay rates as the norm. Thus local rates were very hard to negotiate, although by 1977 Crosse & Blackwell was paying bonuses above this level and the union in turn was able to use the level set by governments for wage restraint as their target.

The 1970 dispute 'brought sections of the Peterhead community to the verge of a nervous breakdown and the whole situation was extremely tense', according to the full-time regional official of USDAW. The result was that the union had access to all new employees, management was to check-off union dues at source and some areas, like bonuses and work-study, were to be open for negotiation. In addition there was to be no victimisation and small pay increases (£1.75 for women and £1.25 heavy work allowance for men). USDAW suffered a recent loss of membership at Crosse & Blackwell when the union insisted upon women taking equal pay for equal work with men.

In the course of research we did not check the details of these various disputes in the engineering and food factories. The facts and the chronology are no longer altogether clear to participants and were of little direct interest to us. What is quite clear is that union activity was quite vigorous prior to 1970 and that neither unionisation nor disputes were attributable to oil. Even the fishing industry was experiencing some changes; in 1975 deckhands tried to form a union and the decasualisation of fish processing as a result of fish-freezing was leading to the stabilisation of the workforce in Fraserburgh and a rise in union consciousness — no doubt enhanced by current threats to the industry by restrictions upon herring fishing.

It seems that up to a point oil broke the solidarity of the employers who no longer observed their no-poaching agreements and brought them closer to their own unions in discussing ways of checking labour losses. Crosse & Blackwell, for example, in its period of high labour turnover in 1973–4, called upon USDAW to help and advise them. The company had to subcontract work to Fraserburgh. Both management and workers in local firms reported a sense of frustration with the ways in which incoming companies first offered higher wages than locals and then found ways of avoiding pay legislation. We found a certain unwillingness on the part of incoming firms to accept the competitive handicap of wages legislation and they used bonuses to circumvent it: punctual attendance, living at home (or away), 'heavy' duty, length of service all variously qualified for bonuses. Only one contractor resisted this (and reported other firms). We suspect this was because the firm owned many factories and did not wish to compromise them on wages.

The fact that the 'Social Contract' or wage norms were being broken could be readily established by enquiry by the unions. But there was also institutionalised means for discussing this when unions on the construction sites and the local unions met. How far did the unions come together? Plainly USDAW would not experience an influx; canteen and camp labour was recruited locally and USDAW was able to unionise site canteens and one camp. The major incoming unions were UCATT, EETPU and AUEW. UCATT, the construction union, was relatively unable to gain much strength locally from an influx of members to the sites because of the high turnover of construction workers. Many maintained their membership at home, thus most of the construction workers from the Glasgow area in Company Y (on p. 111) were members of Glasgow branches of UCATT. Those who did join locally would move with the others and the UCATT ex-local secretary found himself dealing with a stream of arrears letters for members who had long since left the area. This, he said, 'is no great asset to the union.' None the less UCATT members were being recruited by local shop stewards at Boddam and St Fergus. To some extent this represented fragmentation of the *migrant* labour force rather than an increment to local union strength. Only 3 members turned up for an advertised meeting of UCATT out of an estimated membership of 500 in September 1977.

The EETPU had about 80 or 90 members, mainly at Boddam, and they had virtually no local contacts. The AUEW, however, had established a bi-monthly meeting to co-ordinate the activities of members at Boddam and St Fergus and at the first quarterly meeting of the union in 1977 about 35 members turned up from the Boddam site — where the engineering workforce was just beginning to build up. It was estimated that there were 1200 members of AUEW in Peterhead in early 1977, compared with 345 in January 1970. Members from eight companies, including some of the smaller local companies, were trying to co-ordinate their activities in 1977 when we conducted this research. The 'co-ordination' consisted largely of passing information about wages and conditions, and, crucially, disputes with particular contractors. In this last respect we discovered one bizarre dispute where a subcontractor at St Fergus was trying to escape from penalty clauses in his contract by precipitating a strike. The men did not wish to lose their completion bonuses, so were fighting the management by refusing to strike and therefore needed to warn other workers not to be provoked by the contractors.

By the end of 1978 it was clear that co-ordination would remain at a formal level and that no great solidarity was developing between incoming and local workers in the AUEW. The reason for this underlines the peculiarly dependent status of industrial workers in peripheral areas. The AUEW men on the construction sites were time-served men mainly from the traditional heavy industries of Clydeside. They regarded the Peterhead men, trained by General Motors and Clarksons Tools, as country cousins, green labour, lacking the skills of time-served men. Thus local AUEW union cards were not recognised on the construction site, and local members could only work as labourers or unskilled workers. The convener at General Motors felt that this 'very west coast type of attitude' was quite misplaced because the time-served men had the skills of dying industries, where men like those of General Motors were the engineers of the future. The country cousins would be in work and on top of their job when the others had long been redundant.

In the present circumstances the incomers had no need of the local union because if they had a problem 'they'd sort it out at home' in Renfrew, Paisley or elsewhere on the west coast. The local officials meanwhile were less concerned with relations with incomers than the effects of percentage pay increases widening differentials and the possibility of technological redundancy. On this latter point the officers seemed very well informed and their knowledge of technical developments in industry internationally made them seem anything but green rustics.

The AUEW was the largest and best organised union in Peterhead. It was clear that the union would be the basis for the organisation of labour in the district. The Secretary of the Trades Council was a member of the AUEW as were most of the Labour party activists. With increased numbers, in bases, power stations and gas stations, the union leadership might be able to diversify into political activity. At that time all the union officials spent their time entirely upon industrial matters. The incomers seemed most unlikely to make any significant contribution to the power of local labour on the local Labour party.

Conclusions

I

Peterhead has changed; most obviously the town is physically much larger than it was in 1970. Furthermore there are new economic activities in the town, the servicing of offshore activity is the easiest to see, but there are also more jobs, more construction work, more buying and selling in the shops and bars and more business administration taking place in offices and banks. None the less there is one effect oil is not having: it is not leading to the industrialisation or re-industrialisation of Peterhead. No petro-chemical industry, no processing or manufacture based on oil or gas is taking place in the vicinity.

Much of the economic activity stimulated by oil was construction work in the erection of landfall facilities, jetties, warehousing and domestic housing. Large national firms were employed to undertake this work, but they provided jobs for local workers and subcontracts for local building firms. The first phase of construction activities has now been completed. Most notably servicing, supply and fabrication in support of oil and gas production offshore have provided continuing opportunities for local firms. Engineering especially has expanded into this area and the local firms seem to have a secure future although they could come under pressure from offshore operators or service companies who wished to take them over. On balance, however, it will probably continue to be an advantage to the major operators to subcontract maintenance, repairs and fabrication and this offers opportunities for growth by local firms who may be able further to secure their futures by servicing the fishing fleets and the food-processing firms. The need for offshore maintenance is also likely to grow beyond original estimates

when the full effects of the North Sea upon steel structures are eval-
uated.* Thus it seems likely that work will continue in and around
offshore servicing for some years to come. This runs counter to appearances in 1978: traffic was slack in the harbour and BOC had laid off
about 50 per cent of their workforce. Their £2 million development
was in abeyance. But this situation arose from delayed hook-ups offshore
and BOC hoped to recall many of its men when the delays were overcome and the scale of likely maintenance work had been assessed.

Given that Peterhead is not industrialising on the basis of oil but is
nevertheless changing occupationally and economically, is it developing?
If by development we mean achieving autonomous economic growth,
then the answer is unequivocally negative. The dependence of engineering upon oil demonstrates alone the dependent nature of the new
economic activity.

If we look in upon Peterhead from the world of the giant corporation
we see it has a subordinate status, it is a small factor, hardly to be taken
into account in the world-wide strategies of the transnational corporations, large UK companies and the state. Peterhead is a dot on the map,
its value lies in its location, its harbour and the configurations of the
shore line. The presence of people is, to a degree, incidental to those
who wish to occupy the location and use the facilities. This situation is
not entirely new to Peterhead nor wholly due to oil; the other major
resource, fishing, for example, is something that the state seems quite
prepared to use as a bargaining counter in its European Community
strategy.

Peterhead was more a commodity than a community to its users and
this was most strikingly illustrated by land speculation. The sole objective was to make profit by buying cheap and selling dear without
regard to the local consequences. The speculators were 'outsiders' as
parasitic upon incoming developers as they were unconcerned with
local need or sentiment. Similarly the technical decisions involved in
planning major harbour developments were — to the developers' cost —
made without reference to local interests. There was a sense in which
the hectoring and, sometimes, silly *Buchan Observer* was always right:
Peterhead was being acted upon and had no part in the decisions that
most closely affected its future. None of the oil-related economic growth

*C. Skrebowski, 'Is Britain's oil industry on shaky legs? ', *New Scientist*, 16
March 1978, pp. 714–16.

was autonomously controlled from Peterhead nor in the interests of the town alone.

The main activity entailing oil and gas themselves in or near Peterhead was, as we have seen, neither processing nor manufacture, but *trans-shipment*. The raw materials are to be processed and will create (a few) jobs elsewhere. But this cannot be seen simply as the exploitation of Peterhead. The local community cannot eat or use the oil or gas and there is no way in which they can cope with even the most elementary handling of these materials on the basis of local resources.

In order to understand what was happening in Peterhead, as a sociologist, I turned first of all to development theory which has traditionally tried to explain the extent of the failure or success of less-developed societies to industrialise or develop. Central to such sociological theory has been the question of the way in which the resources of less-developed, third-world, countries have been used by the developed countries, and this at the same time seemed to be the key sociological issue in Peterhead.

Underdevelopment in a third-world country entails severe and often long-term exploitation: the production of a cash crop at the expense of local food needs, the displacement of self-sufficient farmers by foreign-owned agribusiness or the extraction of labour in the form of migration. Hunger, poverty and ill-health are the results of underdevelopment, creating an imposed 'cycle of deprivation' that prevents the inhabitants improving their lot. This kind of crude underdevelopment is not to be found in Peterhead in part because the state intervenes to reduce the extremes of deprivation both of social groups and geographical regions. In the Introduction I outlined possible forms of underdevelopment more appropriate to a part of an advanced industrial society. In fact, we found no job destruction, little absolute immiseration, hardly any unemployment and the conditions we described as potential features of underdevelopment were nowhere near as devastating as those found in a third-world country. We detected a widening of absolute economic inequalities and the alteration of a traditional political structure. The former may be one feature of underdevelopment but the latter is largely beneficial. We could not describe the situation in Peterhead in 1977 as 'worse' than in the 1960s when young people left in large numbers to work in Corby, when fishing profits slumped and employment alternatives seemed remote.

What, then, could a sociologist make of Peterhead in the 1970s? Was it just a small town which, on the whole, has done quite well out of developments that were not of its choosing? Was it a location that had lost a spurious autonomy and gained economic activity? At a very simple level one could answer 'Yes' and offer no more elaborate an account of events. But the way in which these changes came about and their long-term consequences are of further interest and so are the theories we use in trying to understand or cope with the changes.

The long-term consequences of oil-related changes in Peterhead were not the original subject of the research, which was concerned with migrant labour. One task, however, was to understand the locality upon which the migrants might be supposed to have an impact. This necessitated studying a number of separate processes together, adopting a holistic approach to a broad topic, none of which fell into neat subdisciplinary definitions, as a precursor of a more detailed study of migrants themselves. We found ourselves considering questions of migration and labour markets, industrial relations, organisation theory, urban and rural sociology, sociology of development and community studies. In each of these areas we found theoretical and ideological debate of a stimulating kind which reinforced our sense of theoretical inadequacy. In attempting to draw all the intellectual threads together we were looking for one theory to which we could relate separate features of the changes: labour migration, occupational change, shift of political power and altered allocations of resources. In spite of their inadequacies and superficial inappropriateness, we had to remain with theories of development and their critics for a level of theory general enough to encompass all that we observed. In doing this we found ourselves looking in upon Peterhead from the outside in order to see the social and economic forces at work upon it, before trying to understand the consequences of these for the local population.

In understanding the role of Peterhead in what I have called the 'world-wide strategies' of companies and the state, we found initially that the work of the Norwegian sociologist Johan Galtung was especially useful and stimulating because he seemed to be proposing a model of the inter*national* division of labour that was equally applicable to inter*regional* questions.*

Galtung was contributing to a debate about dependent development, itself a critique of traditional development theories.† The main weak-

*J. Galtung, 'The European Community and the Developing Countries', Paper No. 77, 1971, Universities Social Sciences Council Conference, Makere.

†See, for example, A. G. Frank, *Sociology of Development and Underdevelopment of Sociology*, Pluto Press, 1971. The best review of the debates,

ness of traditional theories was the division of the world into sectors at different stages of development. Critics stress the one-ness of the world economic system and the interdependence of the parts. The most productive theoretical developments that take account of this have surrounded the idea of 'dependent development'. Dos Santos has described dependent development in the following way: 'When some countries expand through self-impulsion while others, being in a dependent position, can only expand as a reflection of the expansion of the dominant countries' (I. Oxaal *et al.*, *Beyond the Sociology of Development*, Routledge & Kegan Paul, 1975). The word 'dominant' is significant because Dos Santos is not discussing the functional dependence of equal members of a social structure all on the same path to development, but relations of domination and subordination. In Peterhead we find a small town in a subordinate position to powerful economic and political interests. One problem with the notion of dependent development has been the very generality that was its strength for our purposes. We can, after all, readily agree that the world is one system of interaction, it is a kind of conventional wisdom which tells us nothing at all about the structure of social relations between and within particular societies or the consequences of particular social arrangements for specific populations. Galtung advances the analysis by examining dependent development in terms of an international vertical division of processing, of capital and of labour.* In other words he spelt out some of the specific social relations of dependent development.

The division of processing involves the dependent region in low levels of processing and the dominant in higher levels of processing. Autonomous industrialisation in the dependent region is usually precluded because dependent development is geared to the production and export of relatively raw materials, so that the maximum value may be added to them in existing industrial areas near large markets. This was a fact of life in many colonies and is still a fact of life in many ex-colonies. Opportunities for local entrepreneurs are found mainly in servicing this kind of production. This is the Peterhead experience. Galtung suggests, however, that especially polluting production activity *may* none the less be located in a dependent society. We were told that one important

---

although now a little out of date, is J. Hilal's unpublished paper 'Sociology and Underdevelopment'.

*See above, and his *The European Community: A Superpower in the Making* (Allen & Unwin, 1973) in which the analysis is further expanded and elaborated.

factor in Scanitro's choice of Peterhead for perfectly understandable commercial reasons, was that the UK has the least onerous environmental protection laws of the countries comprising the consortium.

The division of capital needs no comment. The UK itself could not provide the finance needed for North Sea oil exploitation and the northeast and Peterhead certainly could not provide any. Northeast capital went into servicing and, through investment trusts, land and housing. Normally profits accrue to capital. Will, then, all the profits migrate, perhaps to where the maximum value is being added to oil, in manufacture? We cannot answer decisively because state intervention is an important factor, and furthermore the government cannot be changed by the direct intervention of foreign companies as is the case in some third-world nations. But the answer for Peterhead is plainly 'Yes'. It is only a question of how far the profits migrate. Neither a Westminster nor an Edinburgh government will give Peterhead priority when large-scale social and economic problems of a politically damaging nature press elsewhere.

Under a dependent division of labour the research, planning and control of production comes from outside and the region provides unskilled, semi-skilled and service workers only. This plainly is the case in Peterhead where virtually all senior management in oil comes from outside the region or from outside the UK.

But this theoretical formulation really only helped us to organise what we already knew about Peterhead. It was easy to tie the 'dependent development' label on the town and so we fell back upon the use of common-sense empirical definitions of the local features of *underdevelopment*, in order to comprehend the details of local changes.

In adopting this approach to the exigencies of research two partly related questions are raised. First, the outcome of almost any sequence of events or economic decisions by oil-related companies depended upon the nature of state intervention. Almost the only unmediated activity was land speculation. The state has no analogue in the structure of international relations central to development theory. Secondly, it is necessary to relate the spatial to the social distribution of goods and disbenefits. This is because a region consists primarily of lines drawn on an administrative map, and providing thereby a convenient basis for policy. But the policies adopted affect groups differentially within the administrative boundaries and the structure of these groups crosses the boundaries. The two questions are related because the state intervenes ostensibly on behalf of 'regions' in a way that has different outcomes for different social groups. Most obviously these outcomes differentially affect people according to their position in the labour market and the market for goods and services; in other words they affect *classes*. Interestingly enough, having abandoned grand theory of the Galtung and Dos Santos kind, we find ourselves immediately returning to questions of *power* which are central to their analyses.

The approach adopted led us to a consideration of the objective of state intervention, the theories underpinning it, the validity of the theories and the possibility of their serving certain interests.

The significant point about the array of grants and loans available to incoming companies is that there is a theory underlying it. The policies described not only rest upon certain theories, they support particular interests. Capital-intensive industry stands to gain the most, even though the policies are ostensibly designed to create employment. A recent report made a similar point with reference to the SDA (Fraser of Allender Institute, *Quarterly Economic Commentary*, vol. 4, no. 2, Oct. 1978). The agency is charged with the pursuit of social objectives, including the generation of employment for example, but has to do so according to the normal commercial criteria of profitability, efficiency and modernity. None of these factors favour the generation of employment as such and they may discriminate against small, new and indigenous firms.

As was said in Chapter 5, 'it is certainly anomalous that the sector most likely to sustain economic growth, long-term employment and the stimulation of small-scale engineering and electronic works does not qualify for development aid in Peterhead.' Obviously oil-related service industries do not have choice of location; they are where the oil industry needs them. Perhaps the logic is: if an industry will come anyhow, why subsidise it? But the same logic could be applied to Scanitro and Shell; they would have been sited in a technically and commercially suitable location in order to carry on a highly profitable activity. Should they therefore have benefited from the inducements offered to encourage manufacturing and processing companies who might otherwise not choose to come to a development area? What is the origin of the apparent anomaly? It may depend in part on a kind of moral judgement that sees manufacture, the making of things, as creating value and therefore good. Services do not create value and are therefore less good. It seems to be a view inherited from Adam Smith. The other source of the anomaly is the equation of manufacture with labour-intensive activity. Historically it may have been true that making things needed many hands and that unemployment has been caused recently by the decline of manufacture. But with capital-intensive production it is decreasingly the case that work can be created by introducing manufacture. Indeed, efficient manufacture and increased productivity frequently demands less labour as currently evidenced by changes in steel production and motorcar manufacture. If the provision of jobs and skills has been a major objective in the northeast then the equation of development with manufacturing with jobs in the present circumstances is likely to subvert that objective because manufacturing and processing in petrochemicals are capital-intensive, not labour-intensive, whilst 'service' activities create jobs, skills and spin-off developments. Service activities will perhaps have a stronger multiplier effect than processing, in pulling more money into the local economy through wages and the creation of extra demand for consumer goods and services. If it is agreed that private economic enterprise should receive state aid in development areas, it seems illogical to exclude the kinds of firms that are likely to operate in Peterhead, unless, that is, one has a defective theory of the processes at work.

The whole 'service industry' concept is meaningless in our context — as empty as the idea of a 'tertiary sector'. From a policy point of view it would be better to drop the distinction entirely. Those responsible for policy at the regional level are already largely aware of the emptiness of the distinction. But of course the particular problems of Peterhead would not make the state alter national policies, and in this sense Peterhead is a peripheral location subject to policies devised to cope with problems of decline and imbalance elsewhere.

The faulty logic for Peterhead and the northeast has now been followed to its appropriate conclusion in the removal of the Grampian Region from development status. This penalises all the non-oil firms who have to compete with oil, and their employees. Furthermore, it makes it more difficult for these firms to modernise or expand. And yet these local non-oil industries will be increasingly important as the significance of oil declines in the 1980s. Once oil is excluded from the calculations, manufacturing is, in fact, declining. Capital in the region is undergoing the same kind of crisis as in the UK at large. The idea of regional capital is, of course, analytically complex. The degree to which and the manner in which regional capitalism (traditionally based upon primary production, food processing and paper) links with the commercial and financial institutions of Scotland and the UK is an important empirical and theoretical problem. But regions, as such, are just lines on maps dividing the country up for management purposes. Administrators, planners, and policy-makers deal with problems on a locational basis, but this does not mean that they are dealing with an entity (called a region) which has its own problems *sui generis*. They treat aspects of wider problems that may only seem to have geographical locations. In the case of Grampian, the problems found in the region have been confused with location so that when oil does well it is assumed that Aberdeen and Grampian are doing well. The extent to which the *location* of problems makes the problems different or peculiar is an empirical problem. Similarly the extent to which social, economic and political relations in particular locations have a discrete and significant autonomy which enables sociologists to use a notion like 'region' for analytical purposes has to be established and not simply assumed.

My theoretical approach throws a different light upon Peterhead than that implicit in state policies. It also had weaknesses. The theories upon which I drew in trying to understand what was happening in Peterhead stressed the uneven economic development of regions and highlighted the *spatial* or geographical distribution of costs and benefits of economic change. Capital has been extracted from both the British regions and colonies overseas without being invested locally or replaced. Northern Ireland and the Scottish midlands developed textiles, shipbuilding, coal, iron and steel at particular phases of the British imperial economy. Now both experience major industrial decline whilst southeast England and the European industrial triangle flourish. The state has, meanwhile, evolved policies to redress the uneven spatial distribution of industrial changes. But has the whole population of southeast England, or Birmingham, so enriched itself meanwhile at the expense of Scotland, Northern Ireland and the English regions that there is now no poverty, no poor housing, no deprivation of any kind in the southeast? The suggestion is absurd.

The decline of one region may be the necessary condition of the development of another, but the costs and benefits are unevenly distributed *socially* within the regions. There are people unemployed, lacking adequate housing and with poor educational opportunities in Strabane, Stonehaven, Spennymoor and Stepney. Fortunes are made in the city of London at the expense of people in all these locations, just as there is a local class in each place which benefits from local deprivation. The unemployed, the poor and the affluent each have common interests that cut across region or locality and have more in common with people like themselves elsewhere than they have with those in better or worse social and economic positions in their own neighbourhood. The extent to which they actually recognise these common interests is highly problematic as is the concept of the 'national interest' looked at in this way.

It is certain *classes* who are the beneficiaries of the relative deprivation of the declining areas or the growth of the growth areas. The state intervenes here too to effect a degree of redistribution between classes, but it does not act in a neutral way. This was obvious in the late 1970s when, first under a Labour government and then when Conservative policies were adopted to shift benefits from labour to capital, to reduce the social wage and redirect revenue spent on social services into private industrial investment – again apparently premised on the notion that services are non-productive drains upon 'real' industry. The crucial point is that an analysis centred upon the causes and effects of events upon a region or locality misses this class aspect, which may be much more significant. Thus parts of the analysis of Peterhead focused upon possible transformations of local classes and the ways in which they articulate with the wider class structure.

Plainly policy affected industrial interests differentially. Prior to the region's loss of development status, however, it was large, transnational corporations that stood to gain most from government policy. This was hardly unexpected given the nature of the development policy and the theories we have suggested underline it, and the state's identity of its interest with that of the oil companies. North Sea oil has enabled the UK to become a net exporter rather than importer of oil, and natural gas will be a substitute for other imported fuels. Gaskin *et al*. (*The Economic Impact of North Sea Oil on Scotland*, HMSO, 1978) reckon that this will entail a balance of payments benefit of up to £5,000 million a year by the early 1980s. Second, oil and gas are sources of taxation revenue to the state worth between £3,000 million and £4,000 million per annum in the early 1980s. The revenues accruing to central government are described by the authors as 'the major benefits from North Sea oil and gas production'. Thus, although the taxation regime brings enormous benefits to the oil industry itself, the major beneficiary in the UK is the state. In this context we can understand why the local problems of Peterhead were as marginal to the state as they were to the oil industry and why the local authorities could therefore be left to cope within the existing and inadequate administrative framework. They even lacked sufficient research staff to help comprehend current changes and provide data to help cope with future change. Even more importantly, given the significance of oil to the state, a case could have been made by either government for sparing Grampian Region from public expenditure cuts because local authority and welfare state spending would be a contribution to production in a key economic sector. But neither party accepts this non-socialist logic at even the simplest level, it seems.

Before the loss of development status for the region in 1979, the incoming multi-nationals gained much from regional policies, especially through financial incentives. Yet they are companies which could afford their North Sea developments without recourse to state aid and which were forced by geography and geology not financial incentives to come to a development area. They attracted very considerable state subsidies because the whole of Scotland was a development area. Thus wherever Scanitro or Shell/Esso located themselves they would have received substantial grants and tax concessions. The most that local employers could have hoped for was a cash grant or loan towards modernisation. This seemed unfair when local employers had weathered economic difficulties in the local community, were 'loyal' to Peterhead and then had to watch a newcomer, who was only after rich pickings and already forcing wage rates up, gain state aid on a big scale. At least this was how locals expressed it during planning enquiries, and it represents a conflict of interest between large oil-based corporations and local employers, who in some cases were branches of international corporations who had been encouraged to come to Peterhead as part of earlier development plans. A more rational development-aid policy would favour only those companies with a 'genuine choice of location' and existing local firms without regard to their industrial 'sector'. In the event 'big' incomers benefited from state policy rather than small and, or, local firms.

The provision of housing in Peterhead can also be seen as a direct subsidy to industry, as it makes the recruitment and accommodation of staff at all levels much easier. It is generally very convenient and cheap, saving companies the need to build and manage their own housing as was the case for nineteenth-century mining entrepreneurs and other industrialists. Similarly, road improvements, harbour reclamation work and the expansion of telephone services can all be seen as contributions by the taxpayer not only to the state's ability to raise revenue but to company profits. Having contributed to the generation of profits, the taxpayer (including the Peterhead taxpayer) has no say where they are invested. Additional costs have been borne by the people of Peterhead in the form of dirt, noise, congestion on the roads, temporary crowding in the schools, but they have no say over the distribution of the benefits for which they have 'paid' thereby. The profits of industry are partly reinvested and partly paid out to shareholders. Most shares are held by other companies. The privately owned shares are held by 7 per cent of the adult population, but 5 per cent of these shareholders own 98 per cent of the shares. We had no evidence that many of these 5 per cent were resident in Peterhead. So the Peterheadian neither controls the use of nor directly benefits from the profits which he has helped create.

In considering the state it would be a mistake to see it as monolithic and omnicompetent. The planning enquiry to consider the building of the Shell/Esso NGL plant showed the corporations and the state to be inefficient and unable to make sense even of the information they had. But the most persistent failure of the state has been its inability to devise for Secretaries of State 'an oil and gas strategy for Scotland . . . and to indicate the role of the Buchan area in that strategy'.

II

Theoretically we find ourselves at something of a loss. The exploration of the problem belongs to a more technical discussion within sociology. Theories of development and underdevelopment deal with the exploitation of less-developed countries which find themselves at the mercy of naked economic forces, or, more concretely, transnational corporations backed by the force of a national and, or, foreign government. Typical of this would be the discussion found in *Beyond the Sociology of Development* edited by Ivor Oxaal *et al*. Discussions of regional policy tend to focus on problems of creating economic growth in a way that redresses the uneven spatial changes in modern capitalism, as in, for example, Stuart Holland's *Capital versus the Regions* (Macmillan, 1976). What we were studying in the north of Scotland and Peterhead in particular was really accidental. If geological formations could be a matter of policy then the state would have chosen to locate oil and gas reserves off the Clyde, the Mersey or the Tyne where there was considerable labour available, industrial and social infrastructure and major 'regional problems'. But oil was actually found in the north North Sea and Peterhead had to adapt to the fact that it was a useful location for some companies, an immediately profitable one for others and a town of strategic economic significance to the government. In saying this we have laid some stress upon the *dependent* status of Peterhead. But Peterhead has always been dependent, it is not and never has been a

self-sufficient island in the UK economy. The only questions are those
of *how* it relates to that economy and it is around answers to these
questions that we use the idea of dependence.

Peterhead is partly dependent upon world market forces, that is
the demand for oil and oil products, and partly upon the state which to
a degree regulates the profit to be made from oil and mediates the social
and economic impact upon Peterhead. The facts of dependence have
been suddenly and quite dramatically demonstrated through oil-related
developments. If nothing else, the myth of independence should have
been destroyed in Peterhead, the population should all now know the
extent to which they are a small part of national and international
economic and political relations. This is hardly a conclusion that needed
sociological study. But if the arrival of oil was 'accidental' some of its
most striking effects were equally contingent and unintended. The most
obvious was outlined in Chapter 6, namely the creation of substantial
housing resources that became available to a hitherto deprived population as a result of miscalculation or unforeseen changes in world
demand for ammonia. Second, the expansion of services (in the general
sense) created work in the town. But labour is not an undifferentiated
commodity, ethnic minorities, migrants, women are all allocated particular roles in the division of labour. The kind of services that developed
in Peterhead were the kind normally staffed by women. Thus women
were a major beneficiary of the occupational changes that took place in
Peterhead. This was not a planned development. Third, because the
town already had an engineering base, local firms were able to develop
rather than incomers being imported by the clients. This kind of development made possible the rise of a new entrepreneurial class who super-
seded the old elite in the town (who had already been undermined by
local government reorganisation). But here we are dealing in person-
alities, not social movements. Only a handful of men responded to what
they saw as a challenge, the opportunities were limited and precise, they
were taken by men backed by an eager lawyer and a banker experienced
in development projects.

These considerations again raise questions of potential conflict,
between classes, men and women, and the old and new economic elite.
On the question of class our conclusion may be clear. The old ruling
group in Peterhead had always acted like a class — we saw this in housing
policy and I often heard it in expressions of quite strong hatred in con-
versation with businessmen and one-time political leaders. The anti-trade
union attitude of Peterhead can be understood not only because trade

unions were some kind of 'threat' to the harbour and a political remedy to low wages but, more importantly, an assertion of autonomy on the part of an otherwise dependent and voiceless working class. This we had to contrast with the attitudes of more 'progressive' corporations which are accustomed to the state meeting the costs of producing and sustaining labour through the education system and welfare state, and who regard trade unions as important instruments in the control of labour. The reasons underlying these differences do not concern us but they present a sharp contrast between the old, indigenous elite and incoming agencies.

The outlook of this old elite is much more in tune with parts of the oil industry that we did not study, but whose activities provided a background to life and work in northeast Scotland. Offshore there were concerted efforts to prevent unionisation and break existing unions; a helicopter company fought a bitter and successful campaign against recognising a pilot's union, and onshore in more isolated locations construction companies successfully fought unions in order to reduce wages and bonuses. This last point is especially significant. The propensity to workplace conflict is affected by the type of industrial undertaking, by its history of unionisation and by its *geographical location* It would be as foolish to pretend that location does not affect conflict any more than to suggest that race and gender do not.

The larger undertakings moving into or near to Peterhead mainly represented those parts of the construction and offshore supply industry that had reached a relatively peaceful accommodation with labour. But it is important to remember that these accommodations are always, to a degree, temporary and contingent. In this sense they, and the engineering factories and Crosse & Blackwell, represented a relatively 'progressive' part of modern industry in comparison with local employers.

None of the old burgh councillors now occupy positions of prominence in the town and the old elite of managers and tradesmen seems to have lost influence. They had little to offer incomers, since capital and decision-making lay outside the town. The rise of the PBA (quite specifically intended to replace the council) was the emergence of the new business elite concerned with profiting from the opportunities offered by oil and with maximising the benefits to the town. This group plainly wished to play an entrepreneurial role as intermediaries between the town and potential incomers. The association lacked statutory authority and access to funds but it did offer the good will of local business and business contacts for incomers. That it was unsuccessful in its intermediary role is mainly a result of the lack of a local economic 'take-off' based on petro-chemicals.

It is quite clear that there had been a succession of power and status in the town. This was also noted by the authors of the Impact Study conducted in the University of Aberdeen. Prominent figures in business and politics in the late 1960s and up to 1972 seemed to have slipped from public view. Most frequently mentioned were the managers of Crosse & Blackwell and General Motors who were said to be much more a power in the land 'until the coming of oil'. Members of the 'old elite' still perform honorific functions at presentations and school prize days.

The implications of this transition are worth spelling out in more detail because they relate to processes taking place on a societal scale and because Peterhead is not simply a new locus for developments in the national and international economy. It is a location in which national class conflicts can be seen to be fought out in a way that was not possible with labour dependent and unorganised.

The small capitalist in Peterhead benefited from low wages and low rates. But, for the incoming oil-related companies operating in Peterhead in a world-wide range of locations and dealing in vastly expensive construction, or capital-intensive production, wages are not the most important factor. They are prepared to pay to get a job done on time and they are relatively uninterested in their influence upon local wage rates. The government, however, wished to encourage development whilst controlling wages. This discriminated against local employers who could only raise wages to compete with incomers during the brief relaxation of wage restraint. Incomers could set their wage levels on arrival but then improve them through the offer of bonuses and overtime or by subterfuge. The government seems to have turned a blind eye to this. Plainly there are exceptions to this rather simple 'rule'. Incoming firms like BOC and ASCo who intend to stay in Peterhead for some years have an interest in reducing wage inflation both to limit their own costs and to reduce the hostility of the local employers alongside whom they have to work. The local small capitalists were accustomed to a dependent workforce, grateful for jobs, unorganised and willing to accept relatively low wages. The local myth of sturdy independence helped sustain their interests. They were anti-trade unions. The incoming firms were accustomed, in Europe and the USA at least, to dealing with organised workers, thus the unionisation of workers offered a further means of control, or at least rational communication with their employees. This was, at the time, very much the view of both major political parties — namely that the unions should control the workforce and their senior officials police the pay policy currently in force. The 'incorporation' of the working class into the institutions of capitalism so that they accept market relations and the logic of capitalism, rather than having them independent and in opposition, would seem to be the desirable outcome for the state. The local capitalists wanted low rates and taxes and were anti-welfare state. The big corporations also prefer low taxes, but by and large accept the logic of taxation when by spending taxes the state is able to support the families that will raise new workers, the schools that will train them and the medical service that will keep them in good health. The welfare state and the nationalised industries provide good service to capitalism. The state taxes as a matter of course and, depending upon its political complexion and the pressures to which it is subject, shifts the costs of producing and sustaining labour either towards labour itself or capital. With the decline of political consensus and the failing of the world economy from the late 1960s onwards, Labour governments have, in fact, cut the 'social wage', namely that part of total household income which comprises the spending or consumption of welfare-state benefits or services. But they also needed to maintain the electoral support of organised labour. Mrs Thatcher's government wished to take this further in order to reduce personal and corporate taxation. But in so doing she was further undermining the economic basis of the compliance of labour and the basis of consensus politics. She was, perhaps, more in tune with Peterhead's small property owners than the incoming corporations — who are none the less not opposed to tax cuts. These examples show the kind of conflicts in which governments are locked.

A similar kind of argument can be developed in the field of unionisation. The management of firms like BOC and ASCo adopts a pro-union stance. Management adopts what Friedman (in *Industry and Labour*, Macmillan, 1977) has called the Responsible Autonomy strategy towards labour. This 'attempts to harness the adaptability of labour power by giving workers leeway and encouraging them to adapt to changing situations in a manner beneficial to the firm. To do this top managers try to win their loyalty, and co-opt their organisations to the firm's ideals (that is, the competitive struggle) ideologically' (p. 78). The firm then needs to devote less effort to direct supervision and can rely upon labour to adapt to changing conditions of work. Management games help to win employees to the firm's ideas and men are offered stable employment and *careers* in return for loyalty. Friedman contrasts this with Direct Control which reduces worker responsibility and uses coercive threats and close supervision. In our research we encountered extreme forms of this offshore and in construction where managers asserted everything from 'these men are animals', 'we employ the dregs' through to militant anti-unionism, and racism by American supervisors. There is, however, a kind of control that falls between the extremes of Autonomy and Direct Control and this we found in the large non-union firms. This is a mixture of paternalism and frontiersman ethic operating in the peculiar culture of 'sturdy independence' of Peterhead and accompanied (perhaps crucially) by high wages. In these firms the boss can be seen working with the men in his overalls and shirtsleeves late at night, and tries to persuade the men to share his entrepreneurial ethic of enriching and improving themselves by taking the individual opportunities provided by oil. Interestingly enough these firms come into more direct contact with the subcontractors drilling and constructing offshore than do the bases which deal directly largely with offshore managers and supply vessels. The engineering and fabrication firms are therefore thoroughly exposed to frontiersman attitudes — which offshore are used to justify pushing men beyond the limits of physical safety and normal limits of industrial supervision. The onshore firm, however, can point to their high wages and ask what further benefits the men could gain from unions.

The unions grew in the sense that the number of union members in the vicinity of Peterhead increased. But the collective power of the unions, and through them the Labour party, did not grow significantly. It would be easy to see this as another contingent factor whereby incoming workers regarded Peterheadians as country cousins. It is, however, rather more complicated than this and again related to society-wide occupational change.

The time-served men from the Clyde have learnt their skills and achieved union membership through apprenticeship, they have a wide range of skills which Kreckel has called general occupational qualifications ('Unequal Opportunity Structure and Labour Market Segmentation', paper to World Congress of Sociology, Uppsala, 1978). But increasingly they seem like a 'traditional' or outmoded sector of the workforce. Not only are their traditional industries declining but, as the Peterhead engineers were ready to point out, their accumulated skills can be transferred to a cassette which programmes a computer or operates a machine worked by a much less skilled worker. The new kind of worker is likely to have been trained specifically for his job by the employer; his skills may constitute what Kreckel called, in contrast to general occupational qualifications 'corporation-specific qualifications' and be largely untransferable. This new worker from the most advanced form of capitalist production is dependent because his skills are not transferable and because he has little control over the skills he deploys. Thus he (or she) stands in at least as great need of a strong union as the worker in the traditional sector — but the latter will wish to defend his economic interests against the 'dilution' by men with different kinds of skill. Thus social forces were at work in Peterhead which prevented the kinds of union solidarity that the local employers feared but they were not forces created by the local employers. The state too has a role in these kinds of conflict, for example, in the encouragement of the development of micro-processor uses that will further reduce both the size and the skill level of labour forces.

Thus inter- and intra-class conflicts are located in Peterhead and made especially salient by the accidental advent of oil. If I were to start my research again I would concentrate on this and the role of the state in representing and mediating these conflicts and in mediating the 'effects of oil'.

Index